FREQUENCY ACQUISITION TECHNIQUES FOR PHASE LOCKED LOOPS

IEEE Press
445 Hoes Lane
Piscataway, NJ 08854

FREQUENCY ACQUISITION TECHNIQUES FOR PHASE LOCKED LOOPS

DANIEL B. TALBOT

IEEE

IEEE PRESS

WILEY

A JOHN WILEY & SONS, INC., PUBLICATION

Cover Image: Courtesy of Daniel B. Talbot

For general information on our other products and services or for technical support, please contact our Customer Care Department within the United States at (800) 762-2974, outside the United States at (317) 572-3993 or fax (317) 572-4002.

Wiley also publishes its books in a variety of electronic formats. Some content that appears in print may not be available in electronic formats. For more information about Wiley products, visit our web site at www.wiley.com.

Library of Congress Cataloging-in-Publication Data:

ISBN: 9781118168103

10 9 8 7 6 5 4 3 2 1

This book is dedicated to our loved ones who have found their final homes.

TOOLKIT

Several useful tools, free simulation software, and illustrations accompany this book as a special bonus on the web at: `http://booksupport.wiley.com`. The reader can thus follow along and run the same simulations described in this book, as well as claim several free tools he/she will find useful throughout an engineering career.

CONTENTS

PREFACE

Many excellent treatises covering phase-locked loops (PLLs) have been published, but to the author's surprise, few books exist covering the frequency acquisition assistance techniques necessary and/or available. A PLL by itself cannot become useful until it has acquired the applied signal's frequency. This acquisition process has been treated with extreme mathematical and/or graphical rigor in the past, mainly by Viterbi and Jet Propulsion Lab (JPL) in publications featuring phase-plane trajectories. However, most of the PLL circuits in other analyses had no explicit assistive circuitry included. Often, a PLL will never reach frequency acquisition (called "capture" in most texts) by itself without explicit assistive circuits (with exception of the overused phase/frequency detector-based approach that has performance drawbacks during settled closed-loop operation that will be identified and explained later in detail).

This book attempts to bridge this information gap. Since mathematical rigor for its own sake can degenerate to intellectual "rigor mortis," the author hopes to offer a treatment of the subject that can be comprehended by both engineer and technician (except for an occasional formula that might be included for essential reasons and even then the book is organized so that the nonmathematician can skip it and move on to the essence of the thought being discussed). A more important reason exists for avoiding excessive mathematical verbosity: often the objective gets forgotten and lost in the minutia; assumptions are too confining or too optimistic or too pessimistic, and the motive behind the complexity is not well stated. For this reason, we have decided to focus on the practical and to simulate actual situations wherever possible, both for clarity, so that the reader can take an active role in altering the assumptions and qualifications of a certain architecture and then rerun the simulation to see what his or her results might be. It is disappointing to be shown a solution that does not fit the problem.

Most of the approaches in this book have been developed through years of experience rather than from articles and books, but where references exist they will be cited. The author believes that it is good to communicate in something besides "mathspeak" (e.g., the English language is quite powerful) and to go directly to the fundamentals. Einstein once stated, "Everything should be made as simple as possible, but not simpler." So get out your 64-bit Divining Rod and 32-bit abacus. The author's intention is to communicate with you, the reader, not snow you, "educate" you, and "intellectually subordinate" you. One can get subordination for free; one need not pay good money for it by buying a math-heavy book in which the author's main accomplishment is to write as esoterically as he possibly can and leave the reader with a feeling of general intellectual inferiority, even if unintentional. It is a hope that clarity trumps verbosity and it remains for you, the reader, to determine whether we succeed.

1

INTRODUCTION

We must begin any treatment of PLL frequency acquisition with a review of the fundamentals of the PLL. Thereafter, the following subjects can be discussed (not necessarily in that order):

(a) The difference between the type I second-order and the type II second-order PLL and why the type II second-order PLL is usually preferred
(b) The mechanism for acquisition in an ordinary unassisted PLL
(c) The multiplying phase detector (numerical multiplier, analog multiplier, RF mixer, or XOR gate used as a multiplier) and effects of hard limiting
(d) The edge-triggered digital phase/frequency detector
(e) The ramp-and-sample phase detector and a technique to exploit its uniqueness to realize wideband capture acquisition assistance
(f) The quadricorrelator, balanced and unbalanced
(g) The Costas PLL, "Frequency Squaring" Carrier Recovery, etc.
(h) The clock and data recovery PLL
(i) PLL considerations for frequency synthesizers
(j) Signal-to-noise ratio (SNR) of the various phase detectors
(k) Short- and long-term phase and frequency settling effects in a PLL
(l) The effect of a phase detector dead zone on phase noise and wander
(m) The reduction or elimination of phase detector dead zone and impact on reference frequency suppression

Frequency Acquisition Techniques for Phase Locked Loops, Daniel B. Talbot.

Published 2012 by John Wiley & Sons, Inc.

(n) Sweeping techniques
(o) Sweep disabling after lock and a self-disabling method
(p) False locking on data or modulation sidebands
(q) Preventing false lock on spurious signals, such as modulation sidebands
(r) Comments on converting an analog to an all-digital architecture
(s) Limiting acceleration in swept acquisition methods
(t) Cycle slipping due to excessive acceleration or modulation within the loop
(u) Killing the quadricorrelator output during final lock to suppress its otherwise added phase noise; using a dead zone versus a switch
(v) DC offsets and noise and the effect on PLL acquisition
(w) Heroic spur suppression using DC offset minimization and brickwall filtering
(x) Brute force methods such as wideband/narrowband PLL bandwidth modes
(y) The frequency ratio detector using counter techniques
(z) Advantages and disadvantages of different assistive techniques and when to use which technique

2

A REVIEW OF PLL FUNDAMENTALS

2.1 WHAT IS A PLL?

A PLL (phase-locked loop) is a circuit that varies a VCO (voltage-controlled oscillator) frequency (hence its phase) relative to that of an input until it matches that of the input signal [1,2]. If the VCO is replaced with a voltage-controlled phase modulator or a voltage-controlled delay element, it is then called a DLL (delay-locked loop) and cannot operate offset in its resting frequency.

The advantage of the VCO is that by changing its frequency relative to that of the other signal, billions of degrees of phase shift over time can be obtained. A DLL cannot do this, and therefore, looks like a completely different animal. By definition, it has no need of frequency acquisition. Its output frequency always matches its input frequency even before the loop is closed or settled. But its phase excursion is extremely limited, often to magnitudes of 360° or under. A VCO is an integrator for phase, whereas the voltage-controlled delay element is zero-order. By that we mean the following. If we apply a step function (a sudden change in DC from voltage A to voltage B) to the VCO input, the VCO output's phase begins to ramp in the direction of the voltage step (which is the frequency step) and phase accumulates. The voltage-controlled delay element's output phase merely "pops" from value A to value B. Hence, a VCO output phase is automatically smoothed compared to that of the voltage-variable delayer.

We will not discuss the DLL any further, since this book is about frequency acquisition methods for phase-locked loops.

Frequency Acquisition Techniques for Phase Locked Loops, Daniel B. Talbot.

Published 2012 by John Wiley & Sons, Inc.

To realize a PLL, the following building blocks ("block diagram" elements) are required:

(1) a phase comparator or detector whose DC output voltage depends on the phase difference between its two inputs
(2) a filter and amplifier to interface the DC voltage from the phase detector with the VCO control-voltage input
(3) a VCO, whose output frequency is directly proportional (or mainly so) to the applied DC control voltage at its input.

The block diagram is shown in Figure 2.1. The phase detector output voltage is called the error signal "*E*(*s*)". The control voltage at the VCO input is called "V_c" and the output phase of the VCO (which is mathematically the integral of its frequency) is called " $Ø_{out}$ " (or "*C*(*s*)" in the figure). The input signal phase is called "$Ø_{in}$" (or "*R*(*s*)" in the figure, which is the notation used by classical control systems [3]). The element between the phase detector output and the VCO input is a lowpass filter, usually first order, and amplifier. That element has a gain-versus-frequency property called its "transfer function," mathematically represented as "*F*(*s*)". (The letter "*s*" represents complex frequency, not to be confused with the Laplace operator for differentiation.) The transfer function *F*(*s*) is NOT generally equal to the closed-loop PLL transfer function, as we shall show. Rather, it is a completely general mathematical expression representing the gain versus frequency properties of whatever we wish the element to be. If we pick an element having too high an order (too many rolloffs), it will be unusable within the PLL and will cause oscillation when the feedback loop is closed. Notice for now that the phase detector is represented as a differencer. This is a highly simplified representation, but one has to begin somewhere. So far, the loop we are describing is called a "linearized" model; that is

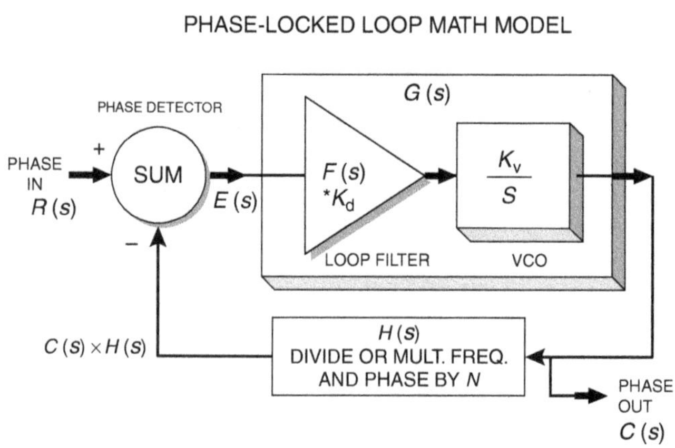

FIGURE 2.1 PLL block diagram.

because the dynamic range (clipping level) of the ideal phase detector is limitless with no overflow or wraparound. Such a PLL would always acquire the applied frequency without any assistance, whatsoever; if we could obtain such a phase detector, you would not need to read this book and this author would not need to write it. But let us press on.

Practical phase detectors (those that can be built) often have wraparound, also called "periodicity." In a nutshell, a periodic phase detector (describing nearly all of them) is one that cannot distinguish between 360 and 720°, for instance. In fact, a human being observing two waveforms on an oscilloscope cannot do any better if he is suddenly presented with the two waveforms without the opportunity of knowing their complete previous history (tracking the phase between them for all negative time). But there is a difference between 360 and 720°. An ideal phase detector would output twice the voltage for 720° as it does for 360. A periodic phase detector, however, outputs the same voltage for 360 and 720° (or perhaps some submultiple or multiple of 360) and repeats its voltage versus phase angle curve. This periodicity may be sine shaped, triangular, sawtooth, and so on. The sine or cosine shape, for example, is a shape that mathematically repeats every multiple of + or -90°. The behavior is comparable to the "rollover" of a car's mechanical odometer when it reaches 100,000 miles. An ideal odometer would not repeat, but would keep incrementing. The shape of the odometer rollover is sawtooth. Mileage accumulates until the maximum (clipping or overflow) amplitude (in this case 100,000 miles), then resets to zero and begins again. Instead of resetting to zero miles after 100,000, if it started to decrement back toward zero, then reach zero before incrementing again, its periodic behavior would be called triangular. An odometer with a 10,000 mile maximum before rollover would be annoying, and one with a 10 mile rollover (overflow) would be exasperating. However, all three would each qualify as an odometer. But one would deal with each very differently. Most practical phase detectors have a very small rollover window analogous to a <<1 mile odometer rollover.

The block marked SUM is the phase detector (in the classical control systems approach, it is actually a differencer). Its conversion gain is K_d that is the voltage produced at its output divided by the phase difference between its two inputs in radians; hence, its units are volts per radian. In analog phase detectors, this conversion gain is highly influenced by signal levels at its two inputs.

Although the detector is typically implemented as a multiplier of waveforms, its function is to create an output proportional to phase difference, so it is properly represented here. Note that this entire block diagram is applicable only to the linear region of the PLL (noncycle-slipping operation[1]), not to the capture behavior prior to lock.

The VCO is shown as having a gain K_v expressed in radians per second per volt. It has an "s" term in its denominator, because it outputs frequency, which is the derivative of phase. Therefore, phase is the integral of frequency, and $1/s$ is the operator for

[1] Cycle slipping refers to the phase detector rollover or clipping behavior we discussed using the example of an odometer.

integration. The VCO is driven by a filter having a transfer function $F(s)$. For a "type 2" second-order PLL this filter has a gain of infinity at 0 Hz (requiring an op-amp).

The full set of equations needed to express the linear model of the PLL is given as follows:

$$\frac{\theta_{\text{out}}}{\theta_{\text{in}}}(s) = \frac{C(s)}{R(s)} = \frac{G(s)}{1 + G(s)H(s)} \tag{2.1}$$

$$H(s) = \frac{1}{N} \text{ for a frequency divider} \tag{2.2}$$

$$H(s) = N \text{ for a frequency multiplier} \tag{2.3}$$

$$G(s) = \frac{K_{\text{d}}K_{\text{v}}F(s)}{s} \tag{2.4}$$

When the transfer function block labeled $F(s)$ is expressed by a zero-order (scalar) expression it results in a first-order PLL overall transfer function, due to the presence of the VCO acting as a single pole at zero frequency.

Thus, if $F(s) = A$, then

$$\frac{\theta_{\text{out}}}{\theta_{in}}(s) = \frac{AK_{\text{d}}K_{\text{v}}}{s + HAK_{\text{d}}K_{\text{v}}} \tag{2.5}$$

If we then substitute $s = j\omega$, we get a lowpass function whose gain at 0 Hz is $1/H$ and whose -3 dB bandwidth is

$$\omega_{-3\,\text{dB}} = HAK_{\text{d}}K_{\text{v}} \tag{2.6}$$

The units for Equation 2.6 are $\sec^{-1}$, interpreted as radians per second. Note several features of this expression. First, the 3 dB bandwidth increases in direct proportion to any of the constants to the right of the equal sign, or in direct proportion to their collective product. Second, the units of K_{d} are volts per radian, the units for K_{v} are radians per second per volt, and the remaining constants have no units. Thus, the product of the terms is indeed inverse seconds.

We see the expression for frequency response of the phase transfer function immediately by inspection, by writing

$$\frac{\Delta\theta_{\text{out}}}{\Delta\theta_{\text{in}}} = \frac{\theta_{\text{out}}}{\theta_{\text{in}}} \tag{2.7}$$

What is the frequency response of the frequency modulation (FM) transfer function? It is simply given by

$$\frac{(\Delta\theta_{\text{out}}/\Delta t)}{(\Delta\theta_{in}/\Delta t)} = \frac{\Delta\theta_{\text{out}}}{\Delta\theta_{\text{in}}} \tag{2.8}$$

Thus, we see that the FM frequency response is the same as the phase modulation (PM) frequency response. This is true in general for any PLL.

2.2 SECOND-ORDER PLL

A limitation of the first-order PLL is the fact that it has a range of phase angles beyond which it loses lock and also fails to capture. This happens when the phase rollover point of the phase detector is encountered, combined with the finite gains of A and K coefficients. On the other hand, a second-order PLL can track over an infinite frequency range, assuming it is type 2. We shall next examine what type one and type two means for a second-order PLL.

2.3 SECOND-ORDER PLL TYPE ONE

A type one (written type I for Roman numeral one) PLL has a lead-lag filter $F(s)$ of the following form:

$$F(s) = \frac{sZ + 1}{sP + 1} \tag{2.9}$$

where $Z =$ lead time constant and $P =$ lag time constant.

Substituting terms, we get the following open-loop PLL transfer function:

$$G(s) = G_0 \frac{K_d K_v (sZ + 1)}{s(sP + 1)} \tag{2.10}$$

Note the presence of a single free "s" term in the denominator. This property makes the PLL a "type I" and the resulting closed-loop transfer function (Eqn. 2.1) is second order.

The type-I PLL closed loop behavior exhibits finite phase error at zero frequency for θ_{in}.

2.4 SECOND-ORDER PLL TYPE TWO

The type-II PLL exploits the characteristics of an operational amplifier to provide infinite gain at zero frequency for the loop filter. The loop filter is thus of the following form:

$$F(s) = \frac{sZ + 1}{sP} \tag{2.11}$$

where again Z = lead time constant and P = lag time constant. But notice this time that the open-loop PLL transfer function is given as:

$$G(s) = \frac{K_d K_v(1 + sZ)}{s^2 P} \tag{2.12}$$

where we now see a denominator containing the product of two free integrations (free "s" terms). When the loop is closed, the result is a PLL whose static phase error at zero frequency is $\{\theta_{in} - (\theta_{out}/H) = 0\}$ at $s = 0$ (i.e., at $\omega = 0$). We have omitted the negative sign multiplying the loop filter equation using the op-amp in inverting mode (required for operation). This negative sign is important when the periodic phase detector has only one slope (sawtooth periodicity), but unimportant for a phase detector having triangular or sinusoidal periodicity. It can be shown that regardless of the sign of the open-loop transfer function, employment of either of the two latter kind of phase detectors will result in a PLL which will slip cycles until it discovers the stable slope providing negative feedback around the loop. The process will be explained a bit later.

2.5 HIGHER-ORDER PLL'S

The need occasionally arises for a PLL of higher than second order, such as when acceleration of the input phase must be tracked (requiring a third-order PLL). Most higher-order loops are not the classical beneficiaries of higher-order terms, that is, the higher order is simply the result of unwanted parasitic poles or sideband suppression filtering (to be briefly discussed a bit later). Such higher order terms are usually arranged to have poles well above the natural loop frequency of ω_n (in the case of a second-order PLL).

2.6 DISTURBANCES

In all cases of order ≥ 2 and type-II PLL or higher, a low-frequency (within the loop's bandwidth) disturbance signal $D(s)$ summed into the VCO input tends to become cancelled by the loop (refer to Fig. 2.2). Proof is left to the reader. This is profound, because it states that a direct current (DC) offset at the VCO input will be perfectly cancelled. Such an offset can represent a voltage offset at the VCO control input, or equivalently a steady-state frequency error. Since we have assumed a linear PLL model, this cancellation is perfect so that any equivalent VCO frequency error, no matter how large, will be cancelled. In other words, since the filter function has infinite gain at DC, the error signal $E(s)$ needs to be merely infinitesimal in order to steer the VCO onto the correct frequency. The two phases at the phase detector input will be such as to produce zero error signal.

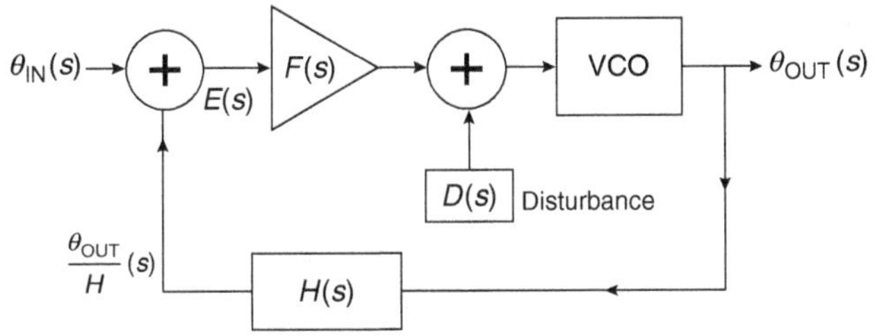

FIGURE 2.2 PLL with disturbance at VCO input.

As the disturbance frequency increases, the cancellation deteriorates. Notice that the disturbance is a VCO-frequency disturbance, not a phase disturbance. We will need to model a phase (noise) disturbance using a summer at the VCO output rather than its input. But VCO phase noise is flat in the FM sense, -6 dB/octave in the phase sense, as we shall see from the later discussion of Leeson's model. Thus, VCO phase noise can indeed be modeled using resistor noise (and 1/*f* shaping if required) appropriately summed into the VCO input.

2.7 FREQUENCY STEERING AND CAPTURE

Thus, we see that the VCO can be steered to the correct frequency from any displacement. Thus, the capture range for our ideal PLL is nearly infinite. This is also true of the PLL model using waveform multiplication for the phase detection, although many cycle slips may be encountered. The way in which an analog multiplier or RF mixer or exclusive-OR gate generates an error signal $E(s)$ is by multiplying the two input waveforms to obtain the trigonometric identity of the product of the two waveforms, which is sinusoidally or triangularly related to the phase difference between them. Harmonics (produced as by-products) are discarded via high-frequency lowpass filtering, usually passive. The two waveforms can be in the RF range whereas the error signal produced is ordinarily in the low-frequency (audio to video) range.

In the case of an exclusive-OR phase detector (which can be modeled simply as an analog multiplier with a comparator or hard limiter at each input), which produces a triangular rollover curve, a frequency error results in an initial cycle slipping beat frequency shown in the example of Figure 2.3.

The same PLL will eventually lock, by producing a weak but essential DC error brought about by second-order distortion [1] of the cycle-slipping beat waveform (see Fig. 2.4). Notice two slopes exist but the PLL will push away if it attempts to reach equilibrium on the wrong slope. Thus, we need not worry about the polarities of the VCO control signal or the phase detector itself.

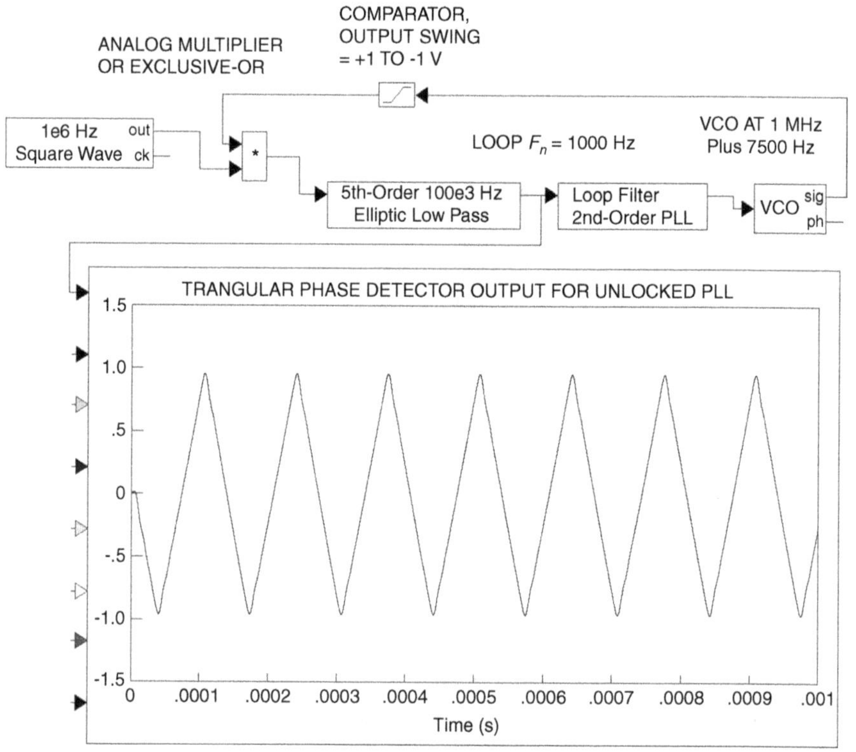

FIGURE 2.3 Unlocked PLL (VCO offset frequency too large) having an exclusive-OR or square-wave input multiplying phase detector.

For a sinusoidally periodic phase detector, the corresponding cycle-slipping beat waveforms prior to and leading toward lock are shown in Figures 2.5 and 2.6.

2.8 EFFECT OF DC OFFSETS OR NOISE PRIOR TO THE LOOP FILTER

For any practical implementation using a periodic phase detector and op-amp-based loop filter, DC offsets and noise may drown out the desired error signal produced by the phase detector. If the VCO is sufficiently off-frequency, the beat waveform second-order distortion may be very weak. This is due to the loop filter rolloff at higher beat frequencies. At some point, the DC value of the phase detector error signal cannot overcome or negate the effects of noise and low-frequency offsets in the op-amp or the phase detector output baseline. When this occurs, two undesirable events take place: (a) the PLL never locks and (b) the offsets steer the VCO in an unlucky direction, eventually slamming its control voltage (tuning signal) into the op-amp rails. For this reason, it is sometimes seen that the infinite-gain property of

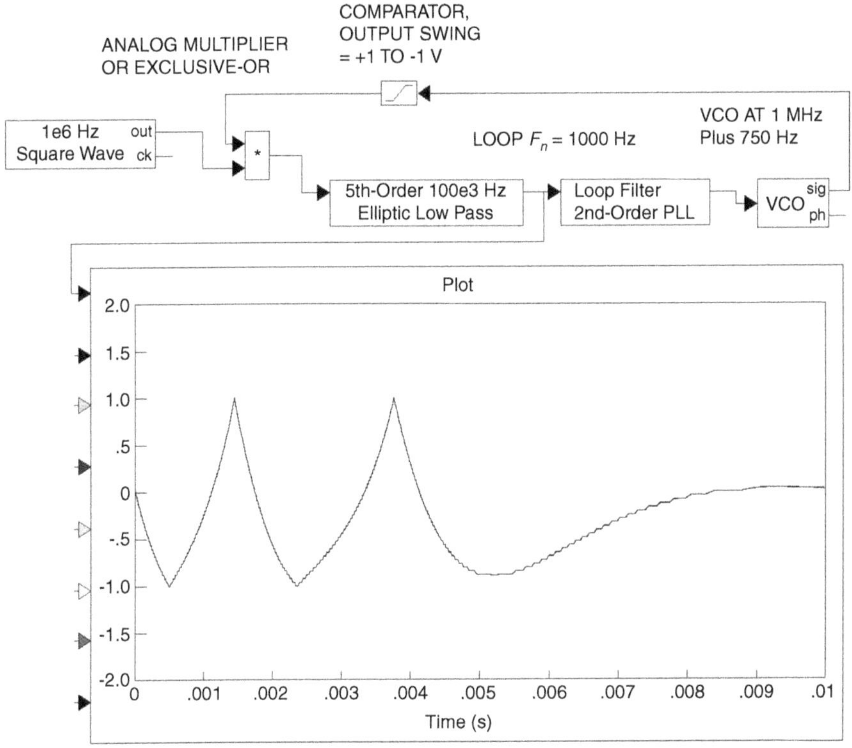

FIGURE 2.4 Locking process for a PLL having an exclusive-OR or a square-wave input multiplying phase detector.

the op-amp loop filter is deliberately spoiled by an added feedback resistor, but this invalidates the PLL as type II and makes it type I. In either case, some external assistance is generally needed to steer the VCO sufficiently close to the correct frequency that the beat waveform can produce a second-order distortion healthy enough to generate the required DC error signal amplitude to overcome the offsets prior to the loop filter (includes op-amp and any other amplifiers in the path from the phase detector to loop filter). The loop filter for type-II operation is often called a "loop integrator."

Perhaps one now realizes why external assistance is required for lock in most type-II PLL's.

That external assistance is the principle topic of this book (although we will cover other related subjects along the way). Some informed readers will no doubt object to the need for assistive circuitry, claiming that some edge-triggering digital techniques for the phase detector eliminate the need. This is true but the use of edge-triggered phase/frequency detectors (see Fig. 2.7) creates performance limitations such as noise aliasing. Thus, edge-triggered phase/frequency detection is capable of mere pedestrian performance in contrast to analog techniques, in which, for example, the

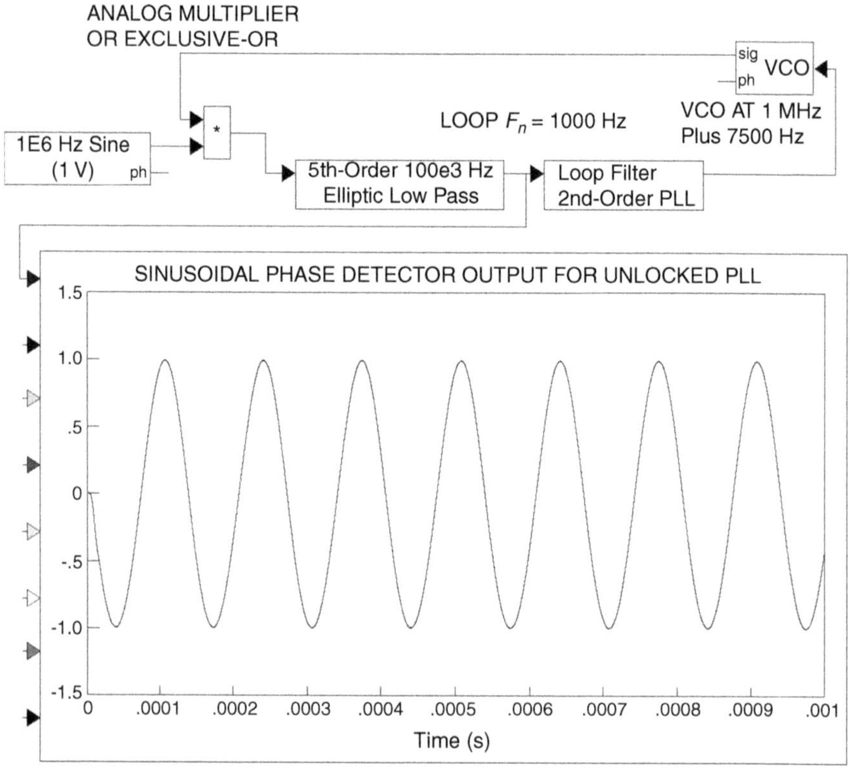

FIGURE 2.5 Unlocked PLL (VCO offset frequency too large) having a sinusoidal input multiplying phase detector.

waveform multiplication can be accomplished without active circuitry (using passive components like diodes), thus, achieving lower phase noise levels. However, if the designer determines via his or her calculations that pedestrian performance is adequate, so be it. Many PLL integrated circuit (IC) designers claim to provide high performance by agonizing over charge pump details, but such fuss is futile. The damage to performance is caused by inescapable noise aliasing and jitter in edge-triggered digital circuitry. Even the digital frequency divider often encountered in the PLL feedback path generates noise aliases. It is recommended that for ultrahigh-performance analog frequency division be considered [4], although programmability may be difficult or impossible. An example of an analog regenerative frequency divider that divides by two is shown in Figure 2.8.

Perhaps the main limitation for the digital circuitry-based PLL is its inability to operate at frequencies as high as an analog-based design. Speed is sacrificed as phase detector circuit complexity increases, especially for sequential logic compared to ring quad diode mixer type phase detectors, which can operate quite well above

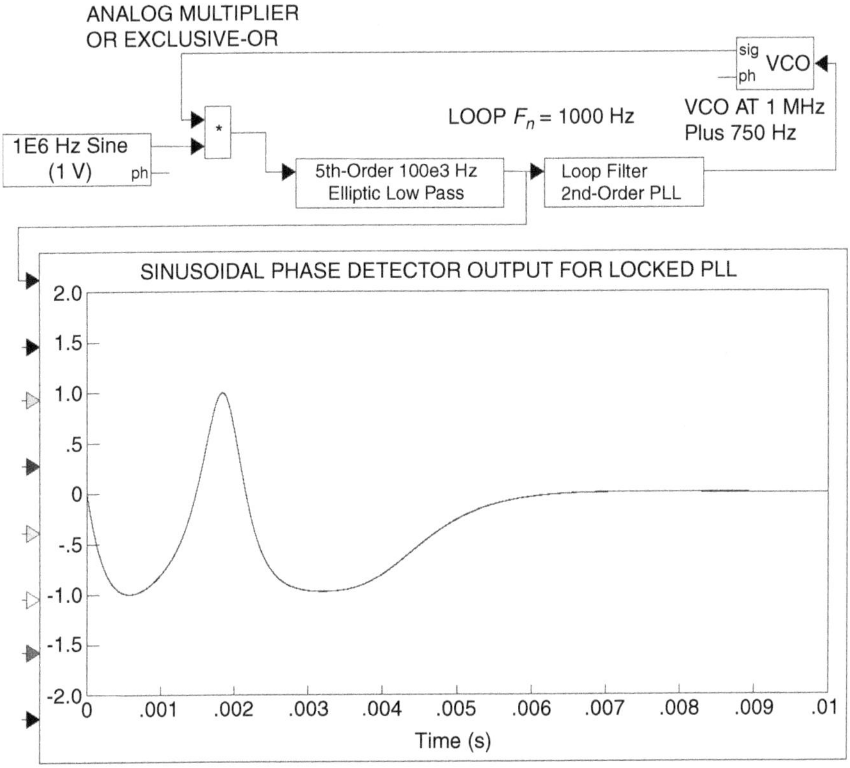

FIGURE 2.6 Locking process (VCO offset frequency ok) for a PLL having a sinusoidal input multiplying phase detector.

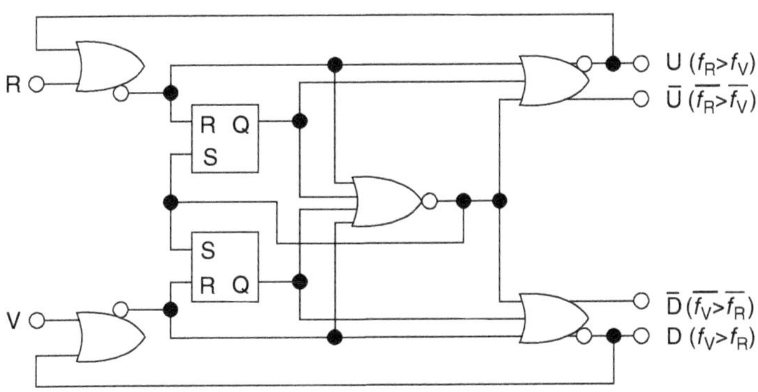

FIGURE 2.7 Typical edge-triggered phase-frequency (type 4) detector.

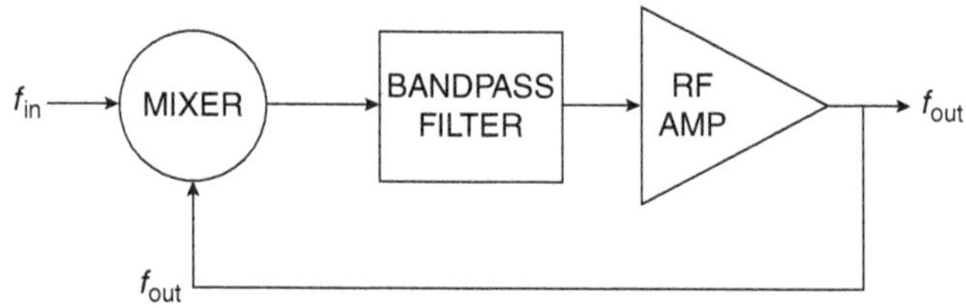

FIGURE 2.8 Analog regenerative frequency divider with $M = 1$ (for division by two).

10 GHz. Also, as digital phase detectors are pushed to operate at very high frequencies, their phase noise increases (their fixed time jitter becomes a greater fraction of their pulse period).

In theory, the analog regenerative frequency divider can work up to the very highest microwave frequencies or down to the lowest audio frequency range (digital prescalers often oscillate when driven by low frequencies because the zero-crossing

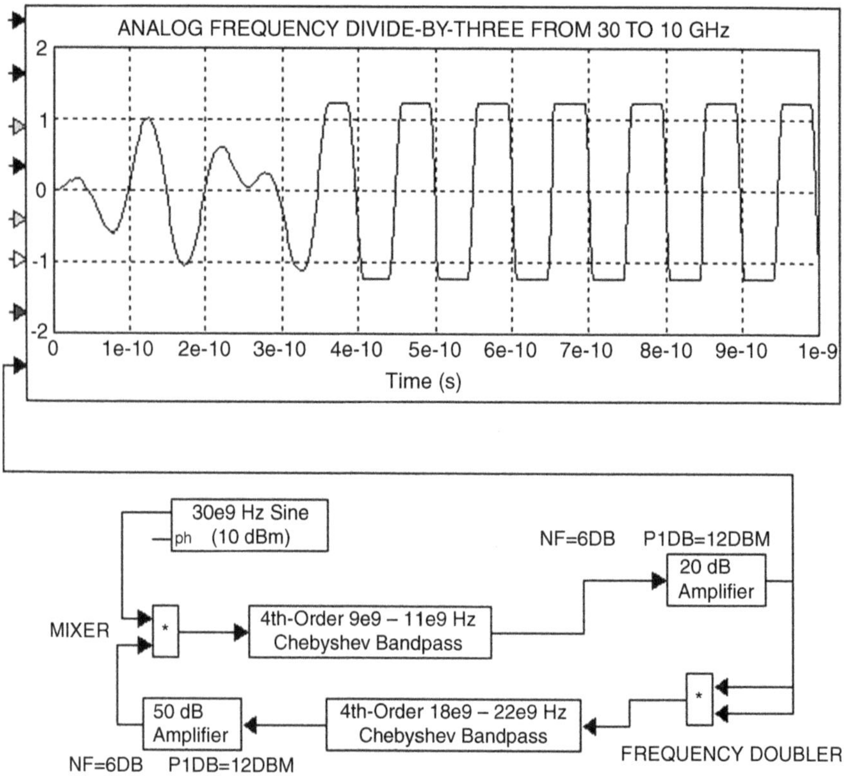

FIGURE 2.9 Analog frequency division by three.

slopes are too shallow to drive the prescaler through its threshold region quickly enough to prevent "parasitic comparator oscillations"—one often sees a low frequency cutoff point named in the data sheet for most prescalers). In Figure 2.9, we illustrate a 30 GHz sinewave fed into the analog divider whose output is forced to be 10 GHz (a division by 3) by virtue of the bandpass filter centered at 10 GHz and a 10 GHz low noise amplifier and a frequency doubler (20 GHz) chain. Note that some noise must be present to start the process, and the amplifier gain must be adequate, and the output builds until a clipping limit is reached, usually determined by the amplifier output saturation, or in the example, by the hard limiting of the RF amplifiers. The frequency multiplier in the feedback chain must be centered at $N-1$ and not at $N+1$ (see Ref. 5). Hence, a feedback frequency doubler works for a divider ratio of 3, but a frequency quadrupler will not work. This is because $N+1$ is not a point of stable equilibrium for the overall scheme.

The 50 dB gain in the frequency multiplier path in Figure 2.9 was used to speed up the simulation for best presentation here. It is important that the equivalent loop gains be amplifying not attenuating, otherwise no regeneration will occur.

2.9 INJECTION-LOCKED OSCILLATIONS

It is also possible to use an injection-locked oscillator as a frequency divider. The original rigorous investigative work on injection-locked oscillators was performed by Adler [6]. Injection locking is desirable in applications where it is intentional; however, it can be unintentional and undesirable in a PLL when it attempts to either lock or influence the main VCO frequency. Designers of PLL's should greatly fear this phenomenon when it is parasitic to normal operation. The paper by Adler is highly recommended reading to understand why. Injection locking can be extremely enthusiastic and potent when the slave and master frequency sources are close to each other in frequency, or their harmonics or subharmonics. The enthusiasm of the injection tendency increases greatly as the two frequencies come nearer each other. For this reason, most PLL's are designed with extreme care regarding leakage amplitudes and frequency schemes. The injection phenomenon poses a tug of war with the phase-lock phenomenon since they each try, in general, to reach equilibrium at incompatible phase angles (unless the leakage phase angle can be favorably established, which is almost impossible especially as frequencies are varied). It has been demonstrated that the effect can be mitigated by increasing the PLL closed-loop bandwidth for a type-II PLL, but in many cases this is neither possible nor allowable.

3

SIMULATING THE PLL LINEAR OPERATION MODE

3.1 LINEAR MODEL

The type II second-order PLL can be modeled as in Figure 3.1 using any SPICE program, but the author has chosen a free software version from Linear Technology Corporation © called LTspice IV (included in the free toolkit for this book).

The voltage source V1 in Figure 3.1 is an AC sweep of unity volts, representing unity radians of phase. Its signal feeds the (input of the phase detector, whose gain (volts per radian) is K_{d}. The phase detector output feeds a resistor R1 connected to the inverting input of an ideal op-amp OA1. Feedback around OA1 is supplied by series network R2 and C1. R2 sets damping for the loop. A PLL exhibiting unity damping factor and $\omega_n/2\pi = f_n = 100$ kHz was chosen for simulation. The VCO chosen has a tuning sensitivity of $K_v = 1$ MHz/volt. This VCO is modeled as a simple integrator (using an ideal op-amp) scaled to $K_v(\mathrm{Hz/volt}) = 1/(2\pi R_3 C_2)$. The values for the resistors and capacitors were the result of calculations automated by the free tool included with this book whose file name is "pllsynth.exe." In that tool, RIN = R1, CF = C1, and RF = R2 are calculated for the user's inputs of K_v, K_d, N, f_n and damping. (*Note*: $f_n = \omega_n/(2\pi)$ and K_v is expressed in Hz per volt for the tool.) The feedback chosen for the example was a divide-by-10, thus H = 0.1, modeled as VCVS (voltage-controlled voltage-source) "FeedbackFactorH." The closed loop magnitude (in deciBels) and phase response versus frequency are shown in Figure 3.2.

Frequency Acquisition Techniques for Phase Locked Loops, Daniel B. Talbot.

Published 2012 by John Wiley & Sons, Inc.

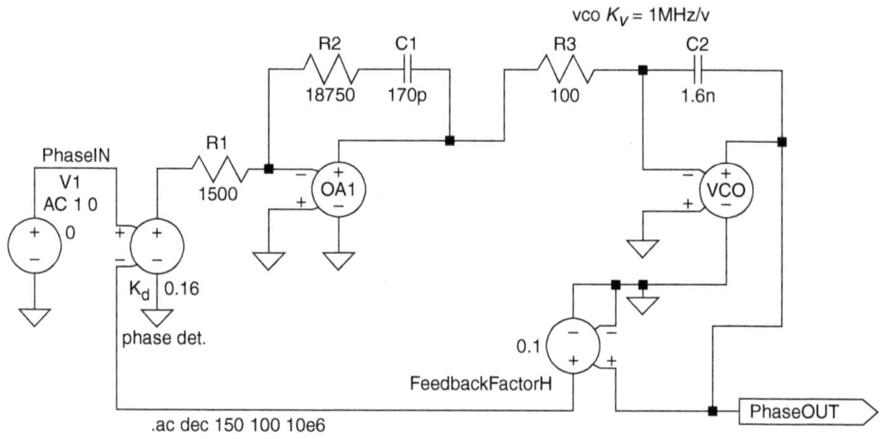

FIGURE 3.1 PLL example using LTspice (frequency response from input to output for $f_n = 100\,\text{kHz}$ and unity damping).

When VCO phase noise is modeled as a disturbance superimposed on the VCO's output, we get the model of Figure 3.3 and the closed loop's response to said phase noise in Figure 3.4.

Notice that the PLL removes phase noise completely near zero Hz and removes it less successfully above that at a rate of 12 dB/octave up to near $f_n = 100\,\text{kHz}$. Thus, the PLL resembles a 12 dB/octave high-pass filter to phase noise from the

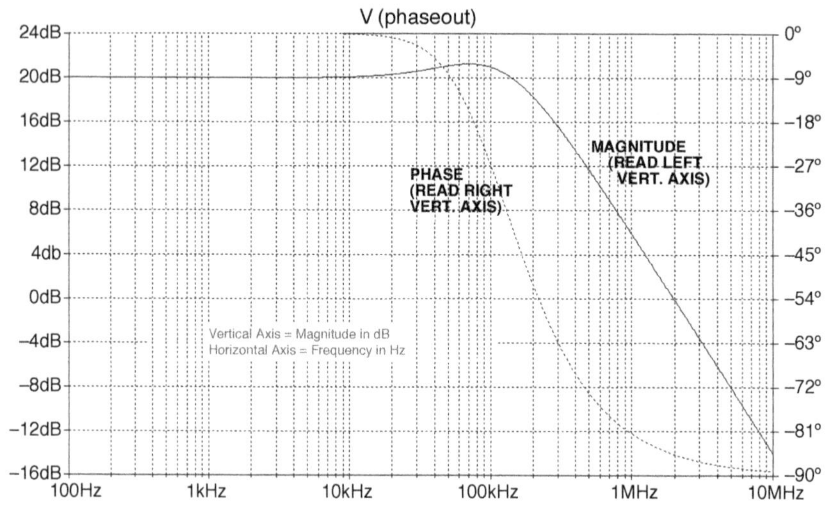

FIGURE 3.2 PLL closed loop frequency response (magnitude in DB and phase in degrees) example.

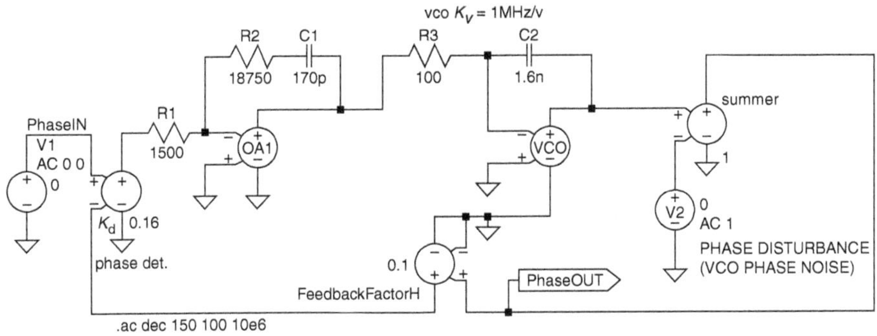

FIGURE 3.3 Modeling PLL response to a VCO phase disturbance.

VCO, but a 6 dB / octave low-pass filter to phase noise superimposed on the PLL reference input θ_{in}.

3.2 A WORD ABOUT DAMPING

The damping factor of a second-order system refers to a relative pole pair location in the s-plane. It is sometimes assumed to have more meaning than it does. For instance, in a normal situation having only a second-order denominator for the transfer

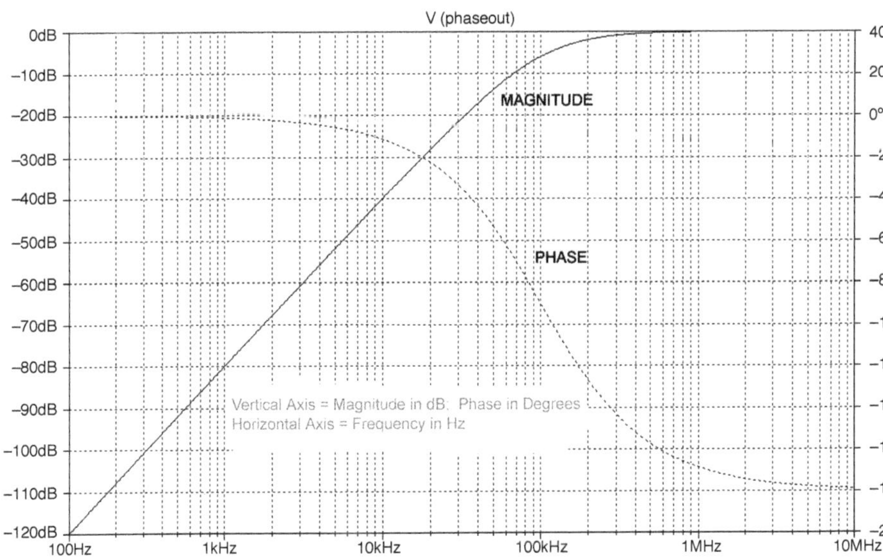

FIGURE 3.4 Response of PLL example to VCO phase noise.

function, a damping factor of 0.707 yields something called "critical damping," and if a time transform is applied to get the step response, there is no overshoot. However, the same damping factor for the denominator pole pair applied to a type II PLL will not exhibit the same step response; in fact, overshoot will be very significant [7]. Thus, one must keep track of one's assumptions. Damping is significant *only* to a pole pair in the transfer function, but numerator zeroes or other denominator poles may render the time response different from that assumed or expected. The best solution is to simulate.

4

SIDEBAND SUPPRESSION FILTERING

4.1 REFERENCE SIDEBANDS AND VCO PUSHING

Often there are extra undesirable signals at the PLL output, caused by the signal energy in the waveform applied to the phase detector, or power supply noise or discrete frequency signals entering the VCO. A practical VCO will exhibit a phenomenon called "pushing," whereby a slight wiggle in its power supply voltage will produce a corresponding wiggle in its output frequency. A realizable VCO may also exhibit "pulling," whereby variations in loading (often caused by driving a mixer whose other input port contains modulation, etc.) convert to variations in the VCO frequency. If these unwanted signal frequencies are well above the loop bandwidth (as usually the case), they will not be suppressed by the loop. Generally, high unwanted signal frequencies appearing to influence the VCO as a component can be addressed (attenuated) only via one of three ways (or a combination): (a) buffering the VCO output so that it doesn't see a time-fluctuating load, (b) regulating the VCO power rail with an adequately low noise regulator (often custom), and (c) increasing the PLL bandwidth to "regulate out" these signal frequencies.

But there is another path in the PLL that can be contaminated. The phase detector has finite isolation of its inputs to its output. Thus, a signal frequency of 2 MHz applied to the phase detector may "leak" to its output, attenuated by some factor, but enter the loop filter and hence the VCO. Because the PLL loop filter has a

Frequency Acquisition Techniques for Phase Locked Loops, Daniel B. Talbot.

Published 2012 by John Wiley & Sons, Inc.

damping resistor that sets a shelf for its frequency response, this phase detector leakage signal (or nonleakage signal sidebands or harmonics) enters the VCO tuning control node rather blatantly. It may cause a problem, or the leakage may be tolerable. However, in most cases it is desired to rid the system of such signal pollution. Often, changing damping and/or loop bandwidth is inadequately helpful. What then? A passive low-pass filter (or an active filter if possible) is often inserted into the phase detector's output path prior to the loop filter, or in some cases, inserted just prior to the VCO tuning input. This passive filter is usually required to have unequal source and load termination, such as when driving a high-impedance VCO tuning line.

4.2 SUPERIORITY OF THE CAUER (OR ELLIPTICAL) FILTER

Such a filter can be designed using the free design tool "fdinstall.exe" in the included toolbox (courtesy of www.aade.com). Figure 4.1 shows a resulting fifth-order Cauer low-pass filter having 50 ohm source impedance and high-impedance output and a 1 MHz cutoff. Figure 4.2 shows the magnitude and phase response of the filter. Why a Cauer filter? The Cauer filter has been much maligned for its abrupt phase nonlinearity near cutoff, but its phase angle near DC is the lowest of all filter categories (for a given stopband effectiveness), and this property is important for minimizing the impact on the PLL's loop phase margin (the Cauer cutoff frequency should be well above the PLL bandwidth and simulation of the overall result using LTspice is recommended). Notice that for this example the phase lag of the low-pass filter is only 20° at 100 kHz. Thus, it is well suited for PLL bandwidths less than that.

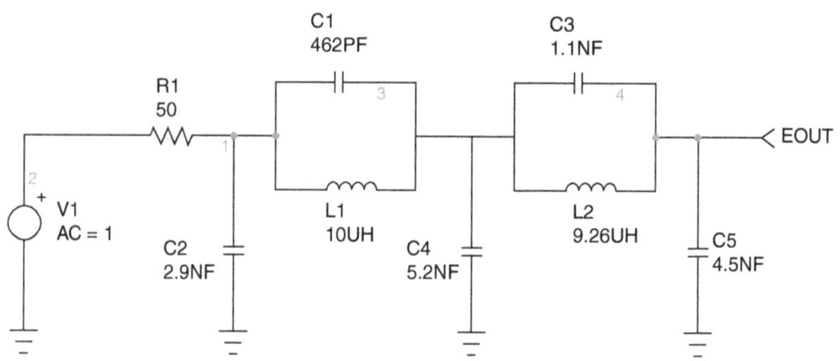

FIGURE 4.1 Cauer low-pass PLL sideband suppression filter example, $N=5$, FC = 1.0 MHz, ripple = 1 dB.

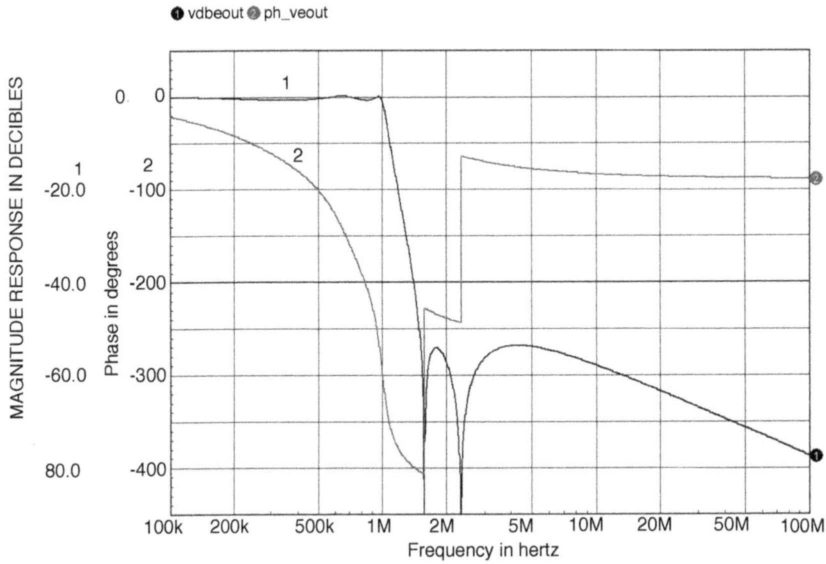

FIGURE. 4.2 Magnitude and phase response of Cauer filter example.

5

PROS AND CONS OF SAMPLED DATA PHASE DETECTION

5.1 WHAT ARE THE FORMS OF SAMPLED DATA PHASE DETECTORS?

Sampled data phase detectors have many forms. First, it might be instructive to know what is *not* a sampled data phase detector component: (a) an ordinary RF mixer, whether single, double, or triple balanced, (b) an analog voltage multiplier, (c) a digital numerical multiplier, (d) an exclusive OR gate, and (e) any device that merely multiplies two waveforms or their numeric equivalents but does not time-sample either waveform.

On the other hand, a sampled data phase detector is one of the following possible types (but not limited to these): (a) a ramp and sample analog phase detector, (b) an RF sampling mixer used as a phase detector, (c) an edge-triggered set-reset flip-flop, (d) an edge-triggered ensemble of flip-flops and gates, and (e) a sample and hold. We will investigate some of these components.

5.2 A. RAMP AND SAMPLE ANALOG PHASE DETECTOR

Refer to Figure 5.1. One input of this component is a trigger that starts a ramp of fixed slope. The other input to this type of phase detector is an impulse train that samples the ramp and holds the result until the next impulse. It is modeled here in VisSim Comm.

Frequency Acquisition Techniques for Phase Locked Loops, Daniel B. Talbot.

Published 2012 by John Wiley & Sons, Inc.

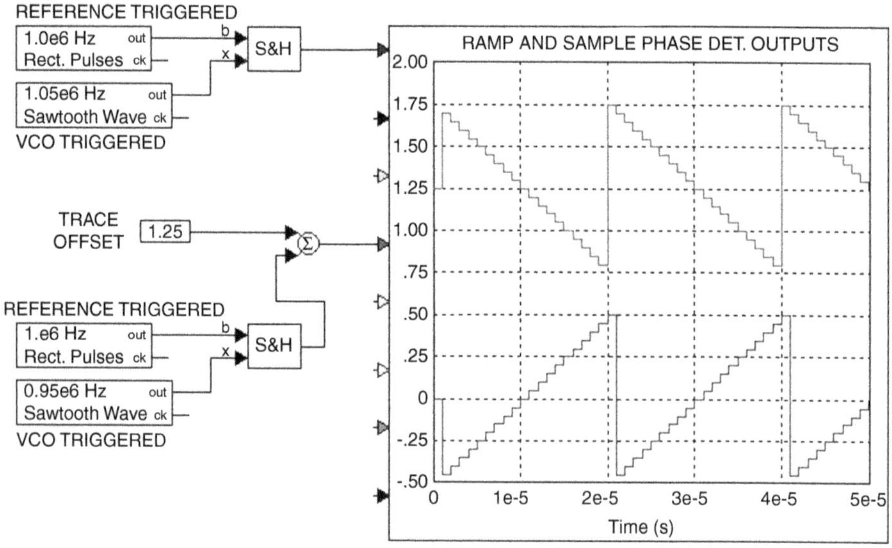

FIGURE 5.1 Ramp and sample phase detector for two cases. Top trace: VCO freq > reference freq; bottom trace: VCO freq < reference freq.

Notice that the flyback (steepest slope) portion of the waveform contains opposite polarity for the case of VCO frequency above and VCO frequency below the phase detector reference signal frequency (VCO in this example includes a VCO divided or multiplied in frequency). Notice also that the phase detector is sawtooth in periodicity. This means it has only one phase slope polarity as far as the PLL is concerned, when operated on-frequency and on-phase (requiring the designer to observe the polarity of each PLL component, unlike the indifference to component polarity permitted by sinusoidal and triangular phase detectors). The PLL will lock when the two phase detector inputs are 180° with respect to each other. The advantage of this type of phase detector is that when the PLL has reached equilibrium, operation will be stationary, that is, no sawtooth waveform will exist at the sample–hold output, only the settled value of DC with no accompanying AC signal sidebands. But any jitter spectrum on either input edge will alias as in any sampled data system. Thus, the problem with aliasing is due to edge triggering, and not restricted to flip-flop-based phase/frequency detectors.

See Figure 5.2 for the zero-frequency error case of static phase between the two phase detector inputs versus output voltage. We see one output slope only between 0 and 360°.

The sawtooth rollover behavior suggests a way to implement a frequency comparator when the PLL is out of lock. See Figure 5.3.

The product of C1 and R1 is chosen to be comparable to the sawtooth flyback time (the sampling aperture) and R2 = R3. The product of R2 and C2 is chosen to be longer than the largest period of the lowest error frequency (typically greater than the PLL loop bandwidth). A buffer may be used to buffer the high impedance of this

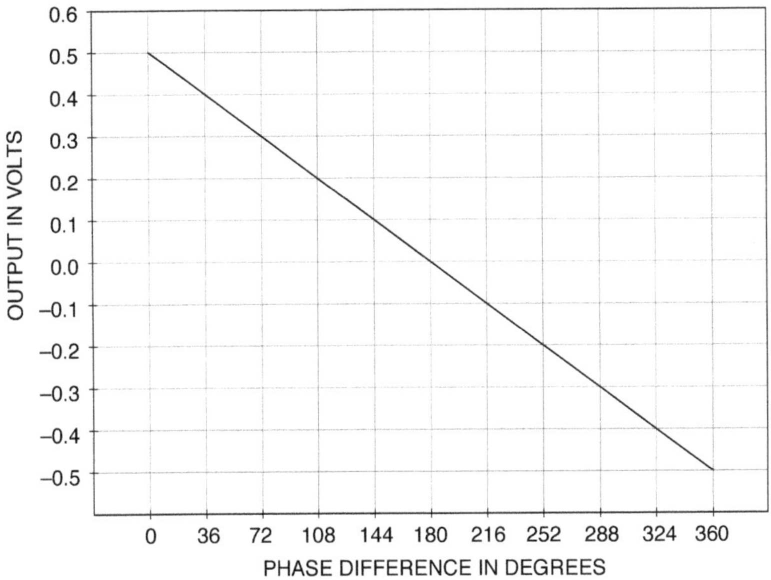

FIGURE 5.2 Ramp and sample phase detector output versus phase error.

network from the loading of a subsequent stage, or the network's output can be fed into a PLL loop node (typically the negative or positive input of the loop integrator, whichever is the appropriate polarity) through a large series resistor to avoid the noise added by a buffer.

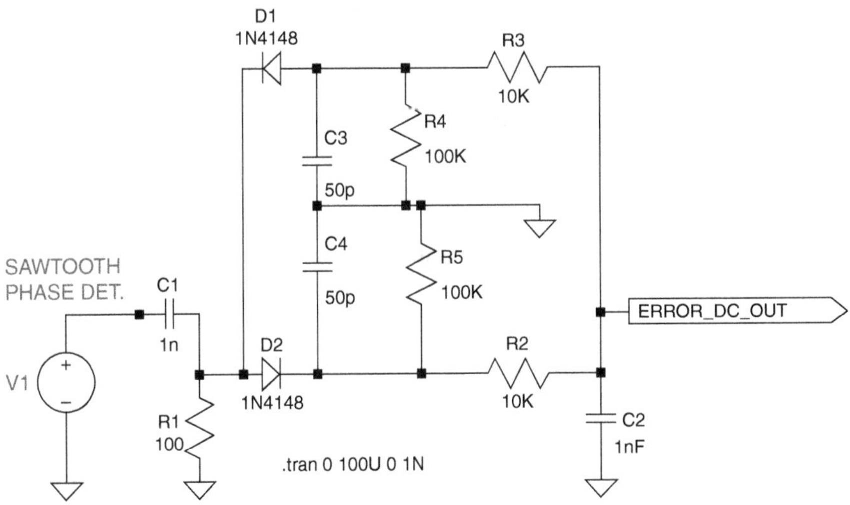

FIGURE 5.3 Method for extracting DC error proportional to frequency offset from a ramp and sample phase detector.

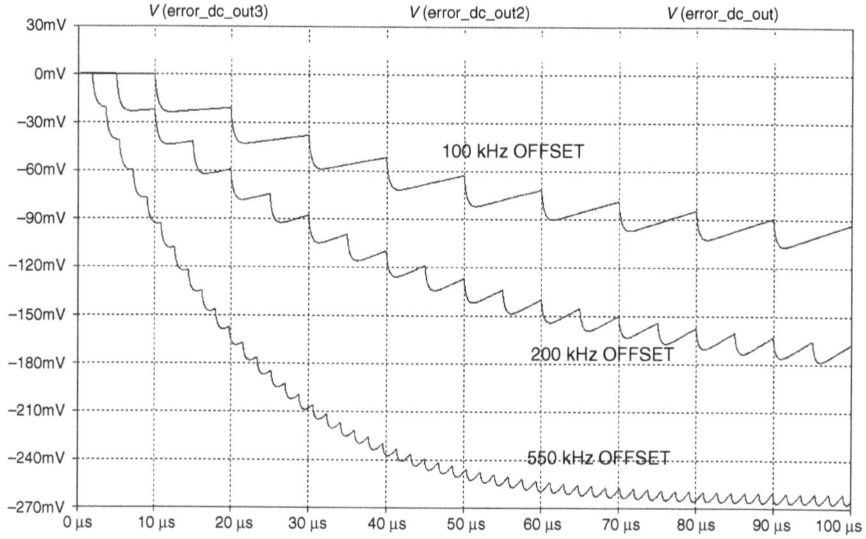

FIGURE 5.4 Freq error detector output for three different freq offsets (for sawtooth phase detector).

The dead zone contributed by the diodes is beneficial, since when loop phase lock occurs, the sawtooth AC waveform from the phase detector ceases to exist, thus disconnecting the diodes from the circuit to insure loop stability in PLL mode.

The network averages the number of flybacks, thus its output is quasi-proportional to frequency error. See Figure 5.4. The sawtooth model assumed 2 V peak-to-peak and 100 ns flyback time. The simulation model can be found in the LTspice directory of this book's web-based toolkit. Note that for higher frequency sawtooths it may be more appropriate to use Schottky diodes for lower barrier and lower capacitance.

5.3 B. THE RF SAMPLING PHASE DETECTOR

Refer to Figure 5.5 . The sampling phase detector exploits the property of a special diode called a step recovery diode (SRD), which, when "pumped" by a strong signal (200–700 mW) whose zero crossings are all equally spaced, exhibits a large spike of voltage resembling an impulse superimposed on the pump signal at every other zero-crossing. Typically, the pump will be close to 100 MHz. The SRD (X2 in the schematic diagram) impulse is about 100–300 ps in width at the HAD (half-amplitude duration) time [8]. The impulse is AC-coupled to a pair of series-connected diodes X3 and X4, via capacitors. X3 and X4 comprise a single balanced mixer, and the new local oscillator ("LO") signal is the impulse train. If the signal marked "FROM_VCO" is one at a frequency N times the LO frequency, it will produce a low-frequency output ("DC ERROR OUT") unless said VCO is at a null in the SRD sine-X-over-X spectrum. Thus, a microwave PLL can be built using no digital feedback division, but which can lock at nearly any N times the LO frequency.

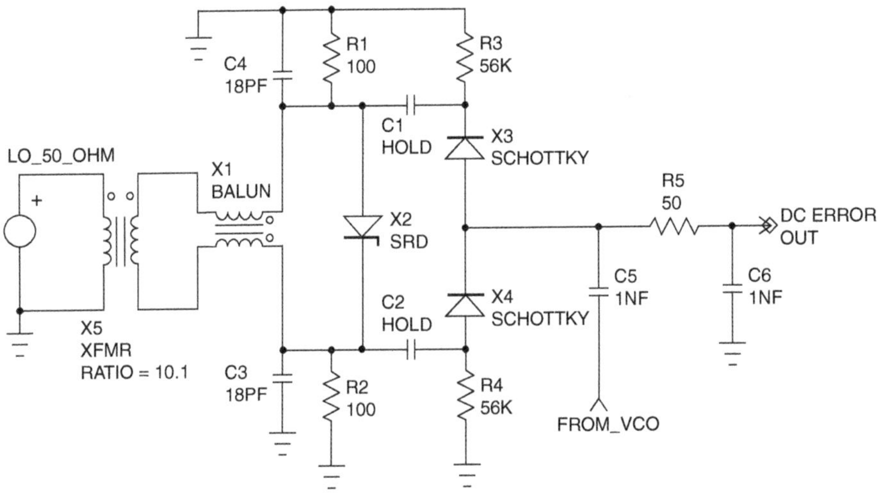

FIGURE 5.5 Sampling RF phase detector using SRD (step recovery diode).

Of course, this means that the VCO must be bounded so that it can lock *only* to the Nth comb spectral harmonic desired and not the $N+M$ or $N–M$ spectral line, any of which is separated from its neighboring comb line by the LO frequency. Recent sampling phase detector technology incorporates MOS switches, capacitor, and non-SRD-based impulse generation in a CMOS integrated circuit. In the SRD-based approach, the capacitors not only provide AC coupling and some rejection of LO pump frequency but also provide an amount of charge storage between impulses. In the more recent technology, the SRD can be eliminated and the diodes replaced with MOSFETs [9].

5.4 C. EDGE-TRIGGERED S-R FLIP-FLOP

It is possible to use a set-reset flip-flop as a phase detector if it is driven by impulses (i.e., if it is edge triggered). A J/K flip may be a more elegant implementation, but the simplicity of the edge-triggered S/R flip-flop lends itself to the explanation of the approach. See Figure 5.6. V1 is a train of positive-going impulses that trigger the SET input of an S/R flip-flop. V2 is a train of similar impulses but shifted in phase from V1.

If the reset impulse occurs soon after the "set" impulse, the flip-flop will assume the "set" state for less duration than when the reset impulses occur later, so the average duty factor will be close to zero. When the reset impulses occur in time just before the "set" impulses then the set time duration will be longer, and hence the duty factor higher. If there is a frequency difference, such as more sets than resets per unit of time, meaning V1 is a higher frequency signal than V2, the S/R flip-flop average output will be greater than 0.5 of full-scale, and if V2 is the higher frequency, there will be more resets than sets per unit time, causing the average output to be less than

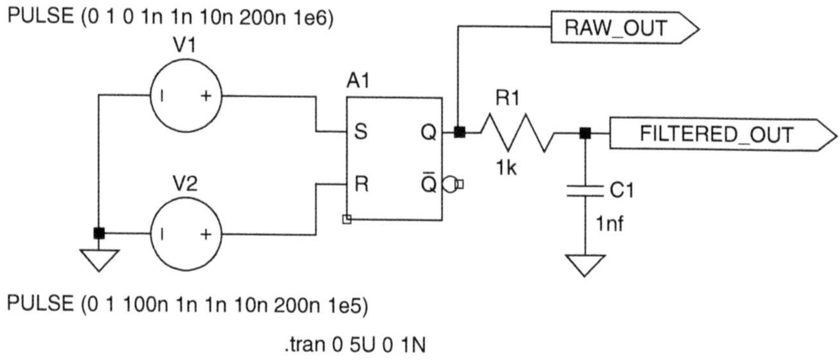

FIGURE 5.6 LTspice model of edge-triggered S/R flip-flop as a phase/frequency detector.

0.5 of full-scale. The LTspice IV simulation model is already in the Wiley toolkit (described in "Preface") for this book.

Like the ramp and sample phase detector, the half-scale output voltage (midway between the two extremes) occurs for 180° phase between V1 and V2 (see Fig. 5.7), but unlike the ramp and sample approach, the output screams with a square wave AC signal at the frequency of operation. So, despite its pros of simplicity, this is a significant con, which places higher demands on the sideband suppression filtering in the PLL. This is also called a type III phase/frequency detector.

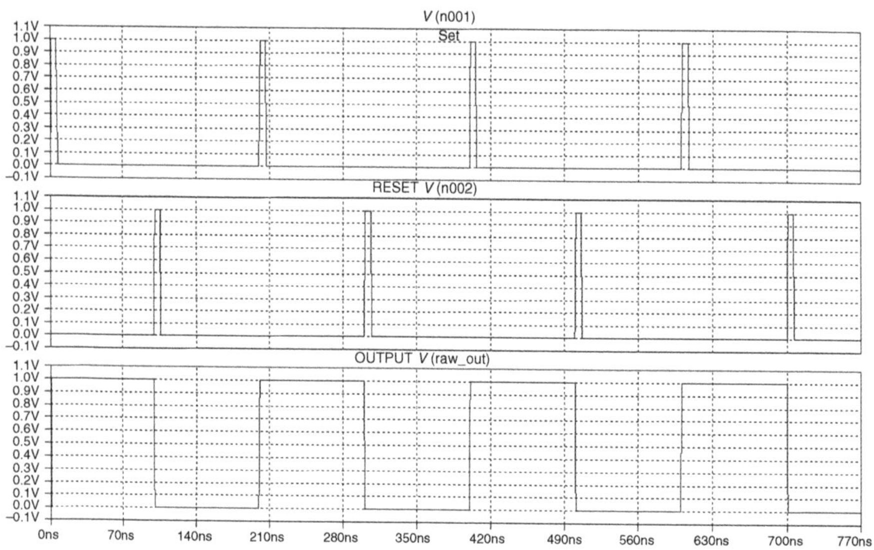

FIGURE 5.7 Waveforms of set/reset Flip-Flop phase detector at a phase error of 180 degrees (midscale).

5.5 D. EDGE-TRIGGERED FLIP-FLOP ENSEMBLE

The multiple flip-flop phase/frequency detector of Figure 2.7, also known as a type IV phase/frequency detector (type II is an exclusive OR gate), improves upon the single edge-triggered S/R flip-flop type III by outputting minimal AC energy at midscale of operation. Thus, the designer can employ a milder amount of sideband suppression. Another benefit is that midscale is encountered for zero-degree phase difference, and the sawtooth output periodicity rolls over at -360 and +360°, thus having twice the linear phase excursion dynamic range. The downsides of this type IV detector are dead zone and, of course, aliasing as in all edge-triggered phase detectors. This type of phase detector is the most commonly used type in integrated circuit PLLs. It is often combined with a charge pump to avoid a subsequent op-amp and its inherent noise. Unfortunately, this results in the employment of only very simple (usually R/C) sideband suppression filters. In order to achieve heroic filtering, the one-port nature of the charge pump–VCO interface must yield to the use of an op-amp or at least an impedance buffer to properly interface to a network like a Cauer filter. In doing so, the charge pump is more of an obstacle than a benefit, which is a pity since so many integrated circuit designers spend so much effort agonizing over its performance.

Another downside common to both the type III and the type IV phase/frequency detectors is the fact that missed edges cause the output to slam to one rail. The analog multiplier, exclusive OR, ramp and sample, and sample–hold phase detectors do not exhibit this offensive behavior, which would spell disaster in a PLL intended for a clock recovery application when the data containing the clock has missed transitions.

5.6 E. SAMPLE AND HOLD AS A PHASE DETECTOR

Another type of phase detector is accomplished using a sample–hold and gating it with an impulse train. Typically, the impulse train might be the edges of a serial data stream from which a clock needs to be extracted. Missing edges simply do not update the sample–hold so it preserves the last known value. That is to say, if the VCO sinewave being sampled remains stationary in phase and frequency, missed edges will have no effect on the PLL using this type of phase detector. That ideal case is shared by the ramp and sample, analog multiplier, and exclusive OR gate types of phase detector. Naturally, none of those ideal cases are perfectly realizable; there will be voltage offsets for the PLL loop filter to contend with, perhaps justifying reversion to a type I PLL. Although the output transfer function resembles that of an analog multiplier, it must be remembered that edge triggering makes this type of detector respond to noise harmonics of the sampling rate, that is, aliasing. The impulse sampling also offers the promise of being able to use an impulse train frequency of $1/N$ times the sinewave being sampled, that is, we can undersample the sinewave from the VCO and obtain a frequency multiplying PLL without the use of a divide-by-N in the PLL feedback path. See Figure 5.8 for the

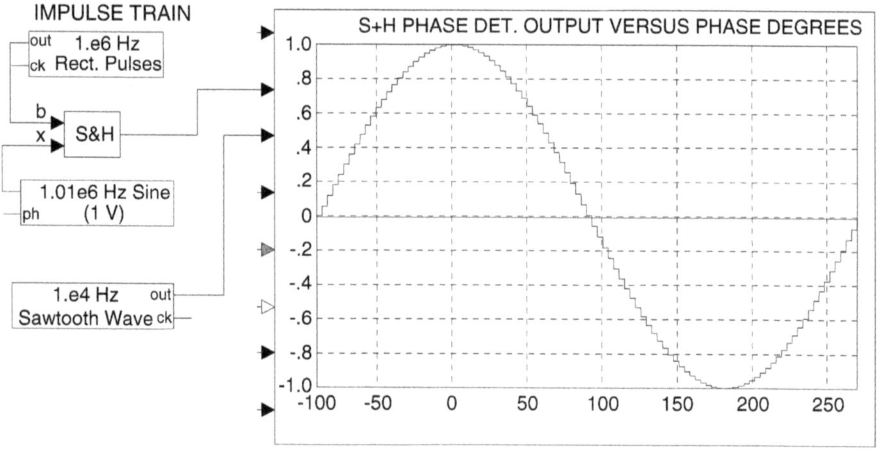

FIGURE 5.8 Sample and hold phase detector with sine periodicity.

phase walk plot; also, see the S + H phase detector VisSim file in the toolkit for this book (VisSim folder, file "S + H_PHASE_DETECTOR.vsm") and experiment with scale factor, speed, and so on.

In most sampled data phase detectors, there is a delay in the interpolated output (the output seen by subsequent stages). It can be shown that even for the edge-triggered digital phase detectors, the delay is one-half the interpulse period. See Figure 5.9 . (Incidentally, a step of phase can be obtained by forcing an impulse into a VCO since a VCO is an integrator for phase.)

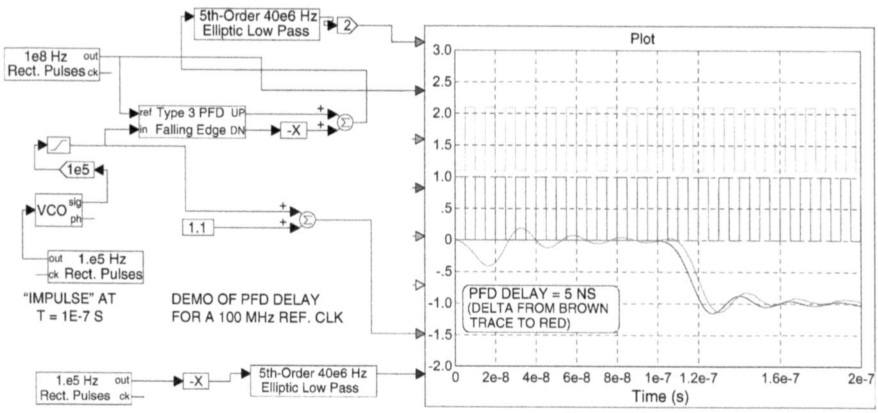

FIGURE 5.9 Phase detector delay (5 ns for a 10 ns interpulse period) = ½ period (most sampled-data phase detector exhibit similar delay). (See the color version of this figure in color plates section.)

6

PHASE COMPRESSION

It is sometimes desired to expand the phase range of a phase detector. This will yield fewer cycle slips in a PLL that employs such an expanded detector. Such a PLL when used to demodulate an FM signal is called an ERPLD (extended range phase-locked demodulator) [11].

The expanded phase detector rollover points are beneficial if there is a lot of noise competing with signal. Occasionally, a noise vector will encircle the signal's phase angle; thus, it is termed "encirclement noise," also called "noise-induced cycle slipping."

The resultant signal-plus-noise vector can exhibit sudden phase flips, which can exceed the rollover points of a periodic phase detector. Remember that a triangular or sinusoidal periodicity has two slopes that cross zero volts. When the signal-plus-noise vector phase suddenly flips so that it lands on the wrong phase detector slope, positive feedback occurs momentarily in the PLL, resulting in "spike noise" or "clicks." By employing a phase modulator in the feedback path around the phase detector, the rollover points can be expanded. See Figure 6.1.

Figure 6.2 shows the VisSim model to simulate the phase compressor of Figure 6.1. (The model file named "PhaseCompressor.vsm" is also included in the toolkit for this book.) The technique can be applied recursively [11]. When a single application of phase compression is used, the result is as shown (in Fig. 6.2). Notice the phase detector rollover points have been expanded almost twofold and most of the curve has a single slope polarity. Also notice that the curve is now almost sawtooth rather than sinusoidal.

Frequency Acquisition Techniques for Phase Locked Loops, Daniel B. Talbot.

Published 2012 by John Wiley & Sons, Inc.

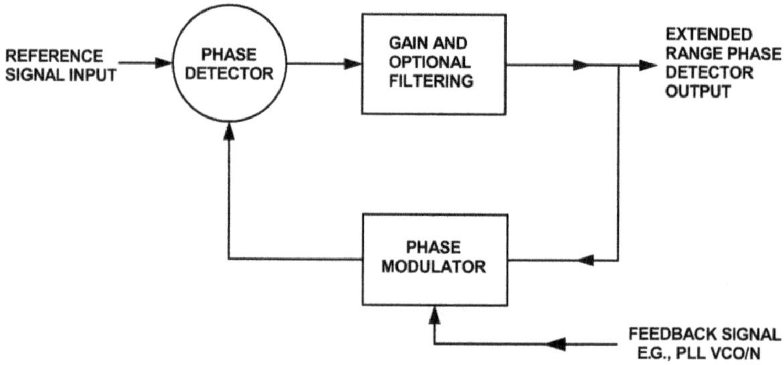

FIGURE 6.1 Example of phase compression (yielding extended range phase detector rollover points).

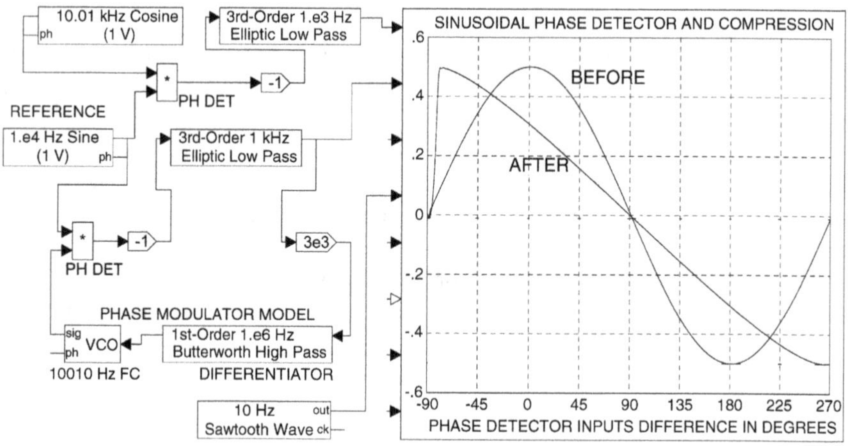

FIGURE 6.2 Phase compression for sinusoidal phase detector.

7

HARD LIMITING OF A SIGNAL PLUS NOISE

In many cases, the signal to be applied to either or both phase detector inputs is accompanied by noise or other smaller signals. In most applications, some form of limiting is present. Even in applications using an RF mixer as a phase detector, there is soft clipping due to the desire to maximize the error signal amplitude compared to DC offsets. This usually requires both mixer inputs to operate at nearly equal power levels, and this condition warps the sinusoidal periodicity into near-triangular, thus behaving much like the exclusive OR phase detector.

A hard limiter is defined as a circuit block that outputs only two levels: $+1$ and -1 or a scaled version of the same. An example is a comparator having one input at ground and the other input connected to the composite signal plus noise. The noise may be incoming from outside the comparator or it may be within the comparator input node(s). The limiter output cannot increase; it is fixed, no matter how large the composite signal becomes. The signal-to-noise ratio is somewhat preserved at the limiter output. Evidence of this is the fact that if we observe the output of the limiter as input SNR increases (the lower trace in Figure 7.1), the limiter's output SNR increases correspondingly. Since this cannot occur due to an increase in signal energy at the limiter output, it must occur because the noise becomes crowded by the increasing signal. In other words, the noise power at the limiter output decreases, even though it remained unchanged at the limiter input. This process is illustrated in Figure 7.1.

It is apparent that the noise is converted to jitter on the signal edges, since the limiter strips away any amplitude modulation. We can explain this very graphically

Frequency Acquisition Techniques for Phase Locked Loops, Daniel B. Talbot.

Published 2012 by John Wiley & Sons, Inc.

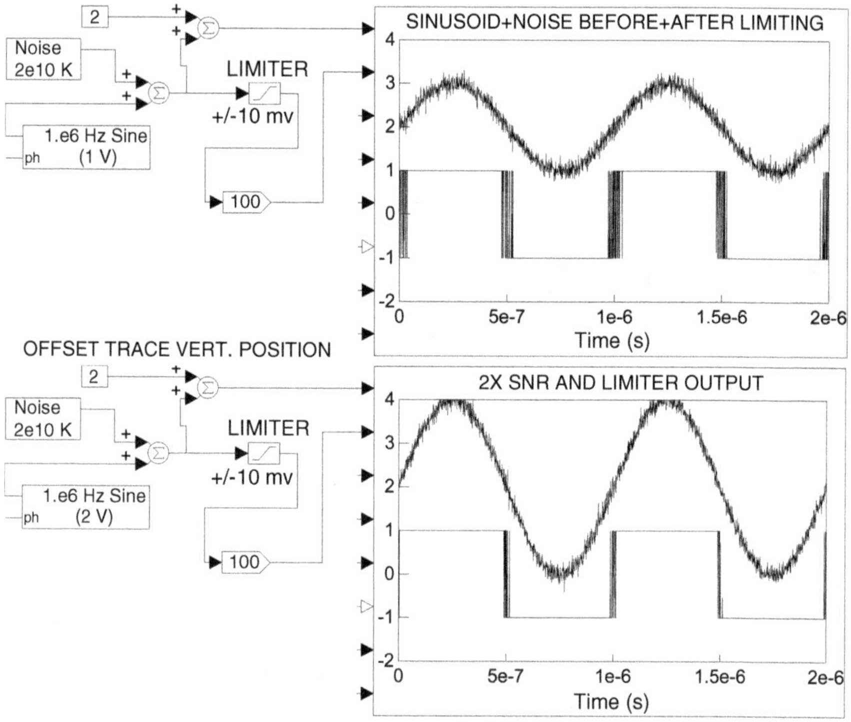

FIGURE 7.1 Two sinusoid amplitudes plus same noise into and out of a hard limiter (comparator).

using vectors. Figure 7.2 shows a main signal of unity amplitude with a secondary signal vector of amplitude β superimposed on it.

This ensemble can be decomposed into an AM signal vector ensemble and a PM signal vector ensemble as shown. The hard limiter strips away the in-phase vector pair of the AM ensemble's counterrotating doublet but leaves the PM vector pair superimposed on the signal's principal vector. Notice that each vector in the pair now contributes $\beta/2$ in units of voltage. Each secondary PM vector now represents exactly 6 dBc less than represented by the original unlimited secondary vector. Thus, hard limiting of a signal accompanied by a single spurious artifact causes a pair of PM sidebands to be created, one at $\omega_c - \omega_{sb}$ and one at $\omega_c + \omega_{sb}$, where $\omega_c =$ frequency of the main signal and $\omega_{sb} =$ frequency of the artifact. Thus, the spectral occupancy is now doubled. This results from standard phase modulation (PM) theory. Taken together, both sidebands add up to radians (interpreted as dBr on a log scale), and the individual PM sidebands are displayed as dBc on a spectrum analyzer.

It is important to realize that noise vectors superimposed on a dominant signal behave identically to what we have just illustrated for sinusoids. Since the same noise vector after hard limiting now becomes a −6 dB lower sideband and a −6 dB upper sideband (both sidebands spawn from the same noise signal and are 100%

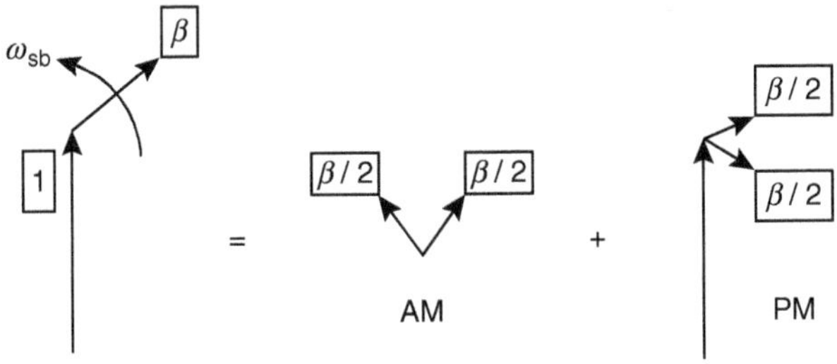

FIGURE 7.2 Signal plus sideband decomposed into AM and PM vectors.

correlated), these two sidebands each contribute half the total noise voltage. That is to say if we are given a single-sideband noise amplitude after hard limiting, the double-sideband energy will be exactly 6 dB higher, not 3 dB as in a power addition [12,13].

In an oscillator, oscillations grow until they are limited in amplitude by a nonlinearity. Van der Pol assumed the equivalent of cubic nonlinearity (followed by Phillip R. Scott of MIT and the use of inverse tangent compression [25]). In any case, the effect of a noise vector in the oscillator tank circuit's passband is to create a double-sideband noise vector pair. Phase modulation of the carrier (the oscillator's center frequency) is thereby the result, and it is called phase noise. Phase noise is multiplicative, that is, one cannot have phase noise in the absence of a signal, since a signal must be present to be modulated by the phase noise. Thus, it is an invalid question to ask what is the phase noise of an ideal resistor! It is unthinkable to see an amplifier as noisy as an oscillator. We will explain shortly.

8

PHASE NOISE AND OTHER SPURIOUS INTERFERERS

Leeson suggested a phase noise model for an oscillator, and his model has been improved upon, but the reader is encouraged to visit the references to his work [12,15]. Figure 8.1 shows the form of phase noise versus frequency for a generalized oscillator, and Figure 8.2 shows a typical phase noise per Hz for a 2.44 GHz SAW VCO. Note a similarity between Leeson's phase noise model and Adler's injection-locking model for an oscillator. Figure 8.3 shows the simulation of an oscillator constructed using a third-order nonlinearity saturating RF amplifier of positive phase angle and a single-tuned (second-order) tank circuit (bandpass filter) providing positive feedback around the oscillator. A disturbance vector is summed with the input to the amplifier. We can picture this disturbance as a sinewave or a single noise frequency component. As one can see, the closer we move the disturbance frequency to the oscillator's operating frequency (carrier), the larger its sideband amplitude becomes, even though we have not increased the disturbance amplitude, only its frequency. This is true for both injection pulling and phase noise.

Leeson showed that the phase noise curve amplitude increases 6 dB/octave (20 dB/decade) as frequency is reduced below a knee in the curve, said knee frequency equal to

$$f_{\text{knee}} = \frac{f_{\text{osc}}}{(2Q)} \tag{8.1}$$

Frequency Acquisition Techniques for Phase Locked Loops, Daniel B. Talbot.

Published 2012 by John Wiley & Sons, Inc.

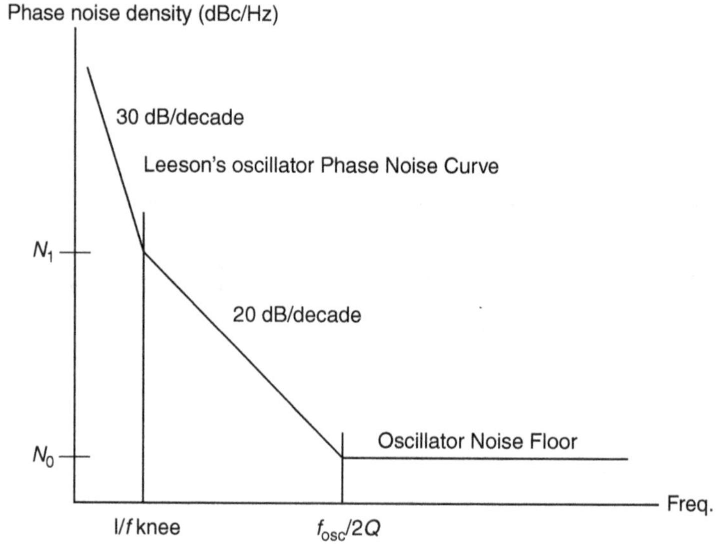

FIGURE 8.1 Typical phase noise density curve for an oscillator.

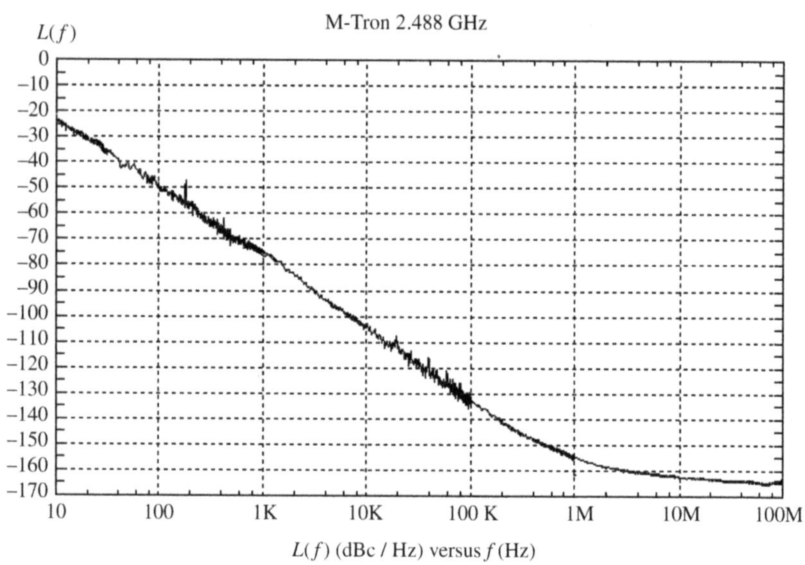

FIGURE 8.2 Actual phase noise of a commercially available VCSO (voltage-controlled surface acoustic wave oscillator).

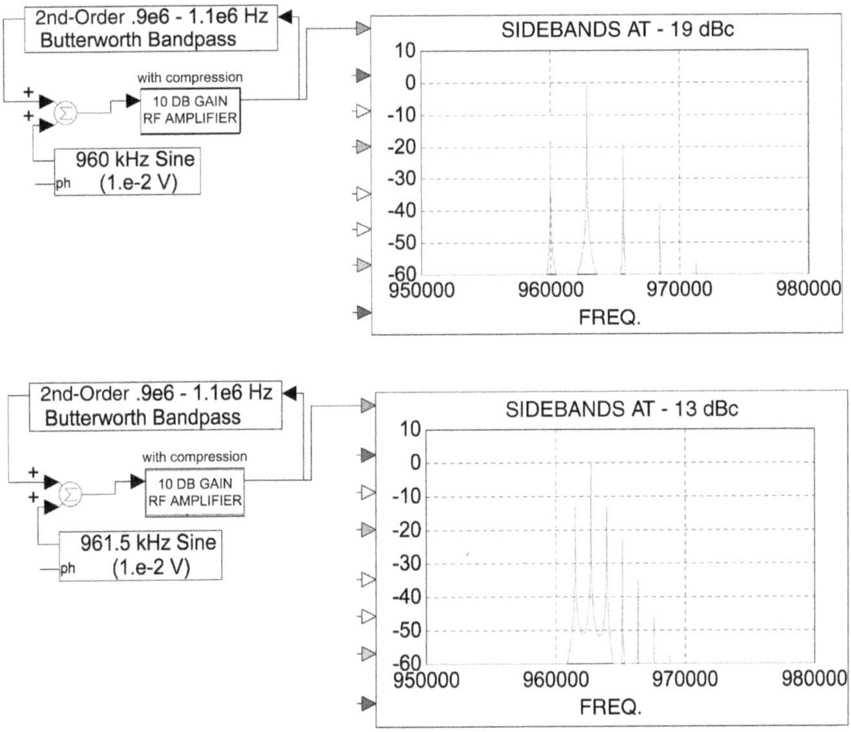

FIGURE 8.3 Illustration of Leeson's phase noise model showing −6 dB/octave slope away from the carrier.

where f_{osc} = frequency of oscillation (or the frequency after multiplication or division); and Q = the Q of the oscillator tank circuit $= f_{osc}/BW$; where BW = the tank bandwidth (or that number after frequency multiplication or division).

Let us take an example. An oscillator generates 1 GHz and does so by multiplying a 10 MHz crystal by 100. Its "Q" is that of the loaded crystal, which is about 50,000. This is the Q whether we consider the oscillator to be a 10 MHz or a 1 GHz unit. Thus, at 1 GHz, the phase noise curve flattens out above a knee of 100 Hz, and rises 6 dB/octave below that knee (assuming a noiseless × 100 multiplier). The knee frequency offset is not noticeably multiplied by frequency multiplication although one might argue that there is some resulting noise modification above the knee due to spectral regrowth. Thus, one only needs to add 40 dB everywhere to the phase noise plot at 10 MHz to yield the new plot at 1 GHz.

It should be pointed out that most curves or tabular data for phase noise are given as dBc (decibels relative to carrier) and are SSB (single-sideband) in nature. Thus, the true total phase noise in dBr (decibels relative to a radian) is 6 dB higher than the SSB noise in dBc.

8.1 THE MECHANISM FOR PHASE NOISE IN AN OSCILLATOR

The key to understanding an oscillator's noise performance begins by first acknowledging Nyquist's criterion for sustained oscillation, namely that the net phase around the loop must equal zero degrees. When a disturbance vector attempts to modify the instantaneous phase angle, the oscillator fights back to reestablish the equilibrium phase (ideally zero). How does it do that? Simply by shifting its instantaneous frequency, and allowing the tank circuit's phase shift at the new frequency to cancel out the disturbance influence. Hence, an oscillator exaggerates its own noise, unlike an amplifier. Remember the equation from frequency modulation that states:

$$\beta = \frac{\Delta f}{f_{\mathrm{m}}} \tag{8.2}$$

where β = radians of peak phase modulation (also called the phase index), Δf = concomitant peak frequency deviation in Hz, f_{m} = modulation rate in Hz (or disturbance or noise frequency).

Equation 8.2 is exact. When one looks at the spectrum of a purely phase-modulated or purely frequency-modulated signal on a spectrum analyzer, the sidebands at $\pm f_{\mathrm{m}}$ around the carrier are each $20 \log (\beta/2)$ relative to the carrier amplitude for $\beta < 0.2$.

For $\beta > 0.2$ harmonic sidebands begin to appear, and at certain greater values of phase index the carrier and first sidebands are highly modified in amplitude according to a Bessel function [13]. A French mathematician once said "Bessel functions are very beautiful functions in spite of their many practical applications." But for phase noise, the amplitude of phase index will be usually $\ll 0.2$ so we need not concern ourselves with Bessel functions (although this author also confesses to liking them very much).

Just a note about Equation 8.2. If we want to build an FM transmitter having the ability to produce as much as 75 kHz peak deviation at a modulating frequency of 20 Hz we would need to produce a peak phase modulation index of 75,000/20 = 3750 radians. A maker of broadcast transmitters did just that. To avoid an automatic frequency control (AFC) loop and to avoid a VCO, the Gates Radio Company manufactured an FM transmitter that used a phase modulator operating on a crystal-controlled carrier. It was called a "serrasoid modulator" invented by J. R. Day in the 1940s [24] and it had difficulty producing the required 3750 radians of peak phase modulation at a 20 Hz rate without distortion. Because of its struggle to generate "equivalent FM" at low audio frequencies, it became a relic of technology even though it eliminated the VCO and AFC loop used in later analog FM transmitters.

8.2 ADDITIVE NOISE IN AN FM CHANNEL AND THE BOWTIE

Since one PLL application is FM demodulation and also since we need frequency steering assistance, we should examine the FM process regarding FM noise and RF additive noise. By FM noise, we mean noise into the control port of a VCO. If we inject flat white noise, it will become flat white FM. Since the open-loop VCO is an integrator, if we demodulate the PM we will see a −6 dB/octave slope versus frequency offset from carrier on a spectrum analyzer (remember the spectrum analyzer displays radians). Thus, Leeson's −20 dB/decade noise curve is equivalent to flat noise into the VCO. This is very significant from a noise-modeling standpoint.

However, if we add flat white noise to the RF path immediately *before* FM demodulation (using a limiter-discriminator or PLL), the recovered baseband spectrum climbs at 6 dB/octave. This is demonstrated in Figure 8.4 . If we could plot negative baseband frequency, the noise spectral density curve would mirror that in Figure 8.4, forming a "bowtie." This is the case for small noise amplitudes [18]. Large amplitudes give rise to encirclement noise. It is important to realize that the behavior for noise is the same as for a sinewave at the noise frequency. Thus, for small magnitudes, the bowtie effect occurs for a swept interfering sinewave in the RF channel. A 6 dB per octave noise rise can be plotted as a parabola for power; thus, additive noise in an RF channel prior to demodulation becomes "parabolic spectral noise power density" at demodulator (frequency discriminator) baseband.

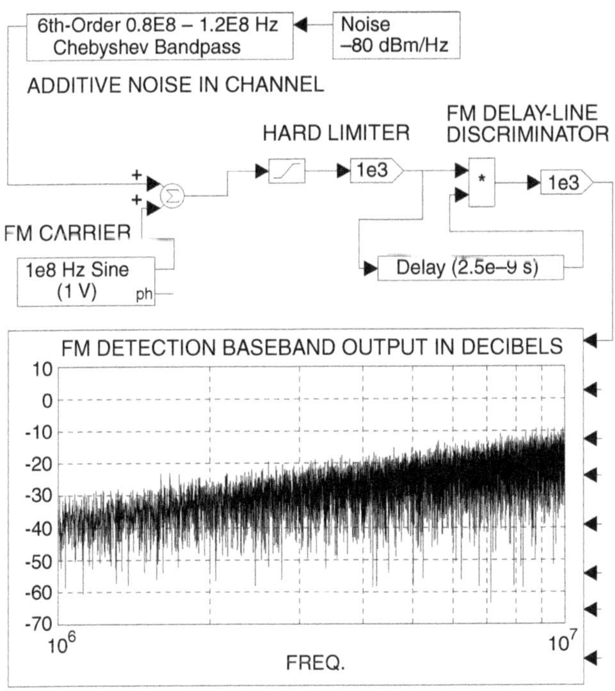

FIGURE 8.4 FM bowtie effect (above the carrier frequency only).

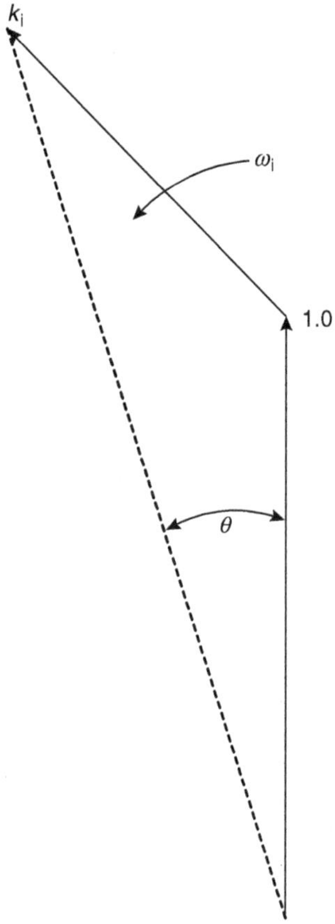

FIGURE 8.5 Vector diagram of RF carrier plus interferer.

It should be clear now that the mathematical analogy for FM detection/demodulation is simple differentiation of phase. Why is this behavior important to know when discussing PLL's in general? It is crucial because the tuning port of the VCO in any PLL is a node exhibiting the output behavior of an FM demodulator. It is also important because we may need a frequency-to-voltage converter (delay-line discriminator or quadricorrelator, and so on) for frequency acquisition.

Figure 8.5 shows the vector ensemble of a unit amplitude carrier vector with an interferer (or additive noise vector at a single frequency) superimposed at a relative frequency of ω_{i} and an amplitude of k_{i} where i denotes interfering signal. Since the vector ensemble's resultant phase angle after hard limiting is given as

$$\theta(t) = \sin^{-1} \frac{k_i \sin \omega_i t}{\sqrt{1 + k_i^2 + 2k_i \cos \omega_i t}} \tag{8.3}$$

and since the output of the subsequent FM detector is directly determined by the derivative of $\theta(t)$, then the recovered baseband waveform is given as

$$\frac{d\theta(t)}{dt} = \theta'(t) = k_i\omega_i \frac{k_i + \cos\omega_i t}{k_i^2 + 2k_i \cos\omega_i t + 1} \tag{8.4}$$

$$\text{Let us set } \omega_i t = \psi \tag{8.5}$$

and then let us take a derivative of $\theta'(t)$ with respect to ψ in order to compute the maximum and minimum of the recovered FM baseband waveform by setting the absolute value of that derivative equal to zero. After a bit more substitution, the maximum and minimum points of the nonsymmetrical resulting waveform are

$$\theta'(t)_{\max} = \frac{k_i\omega_i}{1 - k_i} \tag{8.6}$$

$$\theta'(t)_{\min} = \frac{-k_i\omega_i}{1 + k_i} \tag{8.7}$$

Clearly the beat waveform is nearly symmetrical and sinusoidal for $k_i \ll 1$ and becomes highly nonsymmetrical [17] as $k_i \rightarrow 1$. As $k_i > 1$ the interferer, not the desired carrier, determines the phase velocity of the vector ensemble, and FM detection is centered at this new frequency. This is called the capture effect in FM demodulation [16]. In a PLL, for minimal cycle slipping $k_i \ll 1$ must be true. If not, the resulting clicks will be a function of the PLL's ability to follow acceleration and the phase detector rollover points. When $k_i > 1$ the interferer encircles the carrier vector, and if the interferer is noise, it is termed as encirclement noise. Figure 8.6 shows the result of a 100 MHz 1 V peak carrier summed with a 101 MHz interferer of 0.5 V peak amplitude, said summation hard limited and applied to a delay-line FM detector (discriminator). Notice the asymmetry of the demodulated 1 MHz beat waveform. This simulation is performed using VisSim and is available in the toolkit for this book, mentioned previously. The reader is encouraged to try substituting 0.1 V for the 0.5 V interferer amplitude and observe the much more symmetrical waveform. Or one can change the 101 MHz frequency of the interferer to 100.5 MHz. Notice that unless a nonlinearity is present in the FM detector or after it, the areas enclosed under the waveform positive peaks (down to the 0 baseline) equal the areas enclosed above the negative peaks (up to the 0 baseline). Clipping or distortion in the FM detector postdetection process will disturb that equality, resulting in an apparent baseline shift if the waveform is averaged over a long interval.

8.3 IMPORTANCE OF FM THEORY TO FREQUENCY ACQUISITION

The preceding discussion of noise and interference in an FM channel is of direct interest to us since some techniques for frequency acquisition rely on frequency

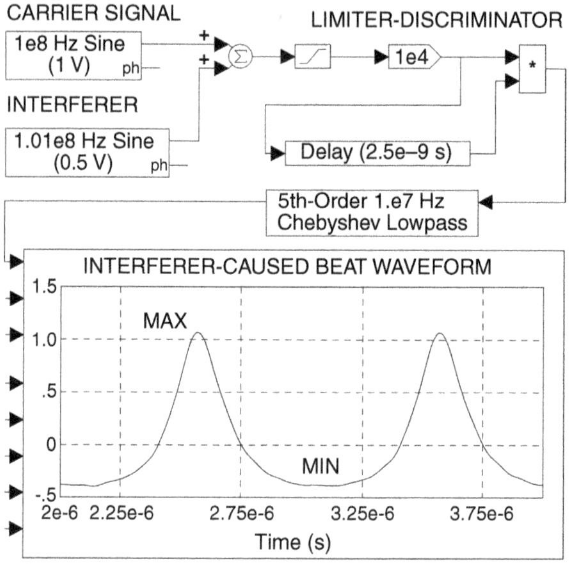

FIGURE 8.6 A recovered FM baseband waveform when interferer is present at significant frequency offset and amplitude.

measurement or comparison. A delay-line frequency discriminator can become an aid to PLL acquisition, although not commonly used in that role. However, a cousin to it is the quadricorrelator, which behaves identically except over a much narrower frequency span but with the benefit of having theoretically perfect 0 V offset at zero frequency error. Virtually all cases of frequency discrimination obey the response to the noise and interference phenomena just discussed. Namely, they all ignore interfering sidebands if we observe their DC average output value, until those spurious signals become large enough to encircle the main carrier, otherwise known as "FM capture." Thus, for almost all situations, the FM detector cannot "false lock" the PLL it is assisting. This is quite a contrast to the PLL itself, which is an easy victim to false locking if left unassisted.

A bit later we will show (a bit to the contrary) that the filters in the quadricorrelator degrade its resemblance to a perfect frequency discriminator in its response to interfering signals, although it can converge to ideal performance when the cutoff frequencies of those filters are made large compared to the frequency error being measured. The reader is strongly urged to remember fully the concepts we have discussed so far in order to follow the later discussion.

9

IMPULSE MODULATION AND NOISE ALIASING

9.1 IMPULSE TRAIN SPECTRUM

Any edge-triggered component, such as the PFD (phase frequency detector) or the sampling or ramp and sample phase detector, possesses a noise performance drawback as previously mentioned. The effective phase noise of such a device climbs at 10 dB/decade as the trigger frequency increases. Why does this occur? Let us attempt one explanation, but first we must lay some groundwork.

Figure 9.1 shows an impulse train at a frequency of 10 MHz and its spectrum. Note the regular spacing of spikes in both the time domain and the frequency domain. In fact, the frequency spectrum resembles a comb whereby each upward-pointing tooth is located at N times 10 MHz, where N is any integer from 1 to infinity.

9.2 SAMPLING PHASE DETECTOR NOISE

We have made the remark that a digital (or analog) edge-sampling phase detector's phase noise is increased 10 log X dB where X is the increase in edge repetition rate. Figure 9.2 substantiates that claim, as well as Ref. [19]. Why is this so? Look at Figure 9.3 . There are a huge number of noise comb "teeth" as seen by the PLL (we have deliberately shrunk the RF bandwidth of each noise comb to emphasize the fact that the PLL gathers only the band of noise that can be converted by aliasing into its passband). As the phase detector operating frequency increases by X, there are X

Frequency Acquisition Techniques for Phase Locked Loops, Daniel B. Talbot.

Published 2012 by John Wiley & Sons, Inc.

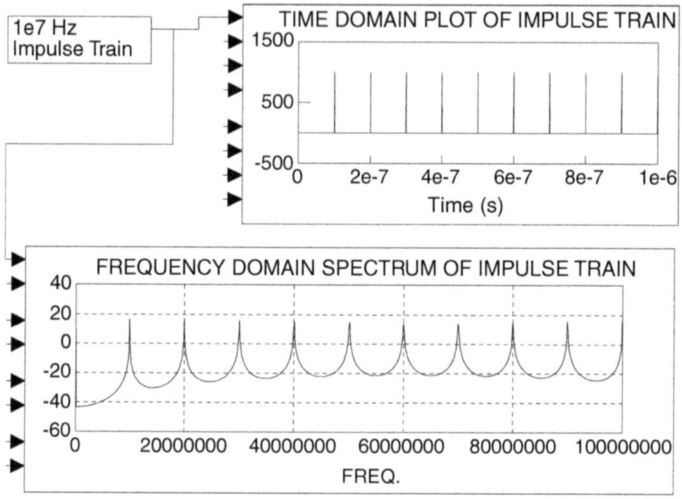

FIGURE 9.1 Comb spectrum property of an impulse train of 100 ns period.

fewer noise clusters (the maximum noise comb tooth frequency is not infinite and is fixed). Thus, were it not for some other noise source, the phase detector aliased noise should be dropping by 10 log X as X increases. But there exists jitter due to "slope intercept" noise (see Figure 9.4) that is fixed (so many picoseconds or nanoseconds or femtoseconds) in the time domain, but as operating frequency increases, the phase

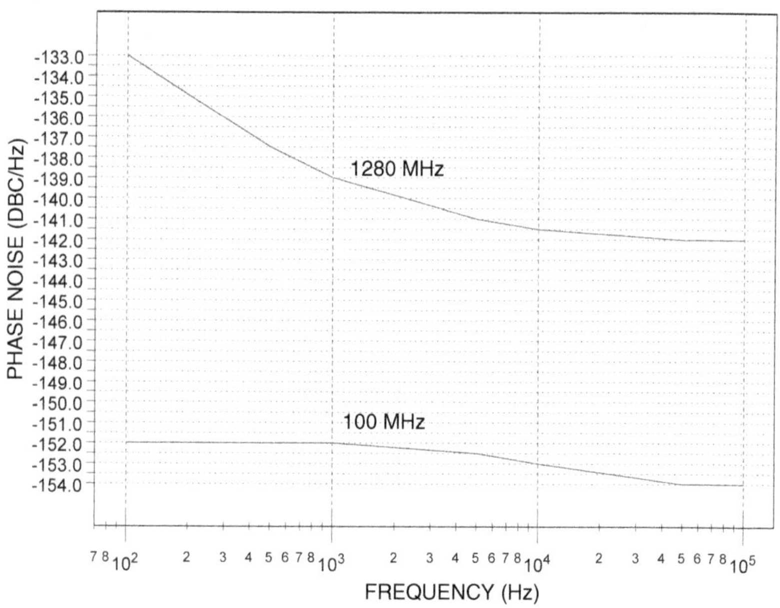

FIGURE 9.2 "Brand-a" actual phase detector measured noise.

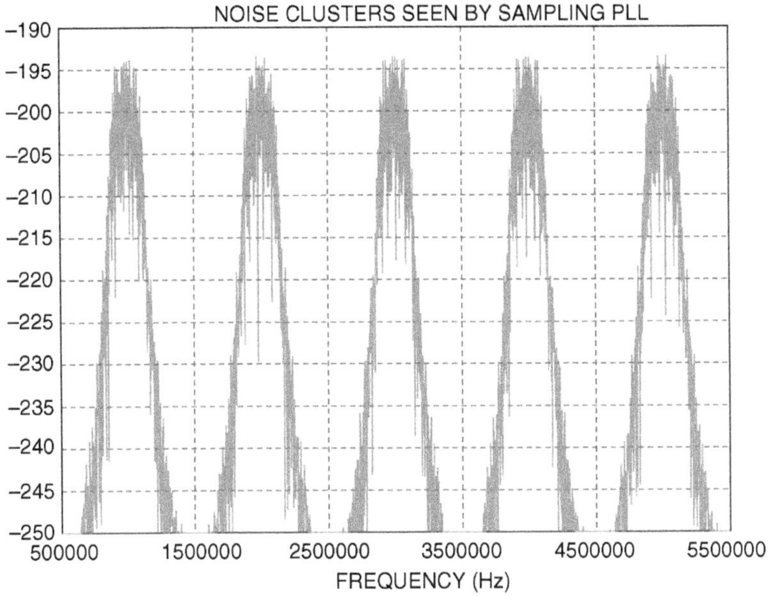

FIGURE 9.3 Noise clumps as viewed by sampling process.

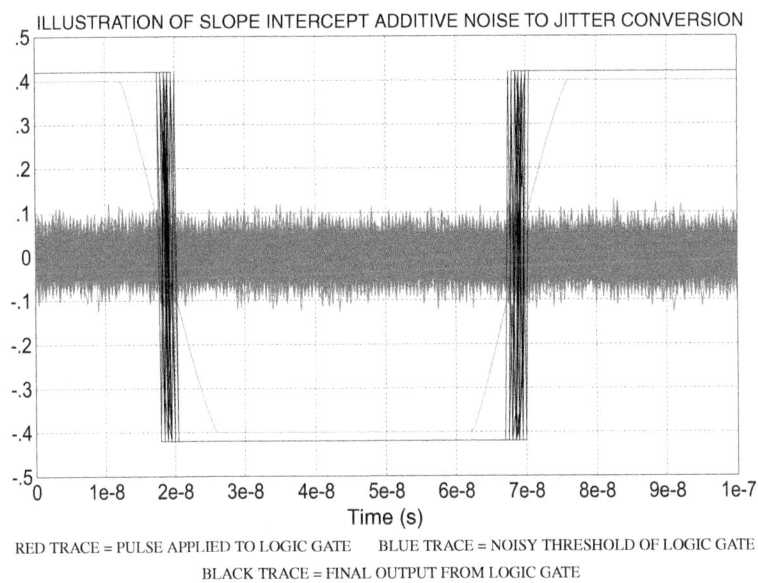

FIGURE 9.4 Illustration of additive white Gaussian noise (AWGN) causing jitter via slope intercept. (*See the color version of this figure in color plates section.*)

jitter represented by this time jitter increases proportionally, that is, 20 log X. Therefore, one mechanism is worsening at 20 log X, another improving at 10 log X, for a net worsening of phase noise of 10 log X. As mentioned, Figure 9.2 proves it.

Experts have defined a term "normalized phase noise" to simplify the computation of phase noise at a specific phase detector operating frequency. The result is a phase noise density in dBm per Hz equal to PND = $\{10 \log (f_{ref} / 1\,\text{Hz}) + Y\}$ dBm/Hz, where Y is some number that depends on semiconductor process. Below are some values for Y as a function of semiconductor process:

Process	Y
CMOS	−230 to −218
SiGe	−230 to −223
TTL	−222
ECL	−200
G_AA_S	−190

These numbers for Y will vary between manufacturers. The noise tends to be flat except near zero Hz.

9.3 SPUR ALIASING

Noise is, of course, not the only entity that can alias down into the PLL's passband (and subsequently into its VCO output). So can signals at discrete parasitic frequencies, called spurs (for spurious). Sampling phase detectors of any type will "alias the universe." For this reason, there are many more opportunities for PLL output spectral contamination when using a sampling phase detector (PFD especially) than when using an analog phase detector. In fact, if the signal being aliased has harmonics, those harmonics may also alias. This is a phenomenon called N-by-M intermodulation. It occurs in RF mixers and A/D and D/A converters to some degree, but nowhere near as vehemently as in sampling phase detectors. In short, the Mth harmonic of a signal cross-talking into the sampler could heterodyne (cross) with the Nth harmonic of the sampling frequency, and this could result in a forest of spurs at the PLL output. To illustrate, we run the program "NXM.EXE" included in this book's toolkit for a phase detector frequency (i.e., sampling frequency) of 100 MHz, a stray cross-talk signal close to 200.05 MHz, and a PLL passband of 1 MHz (RF bandwidth of 2 MHz). We tabulate a portion of the possible results in Figure 9.5 . Note how numerous the cross-products can be. A forest of spurs on the output of a PLL used in a frequency synthesizer or a communications receiver local oscillator may be deadly at −80 dBc, whereas for a clock recovery PLL they might be insignificant. It is highly application dependent. But it can be said unequivocally that the analog phase detector suffers far less from this issue than a PFD or other sampling phase detector.

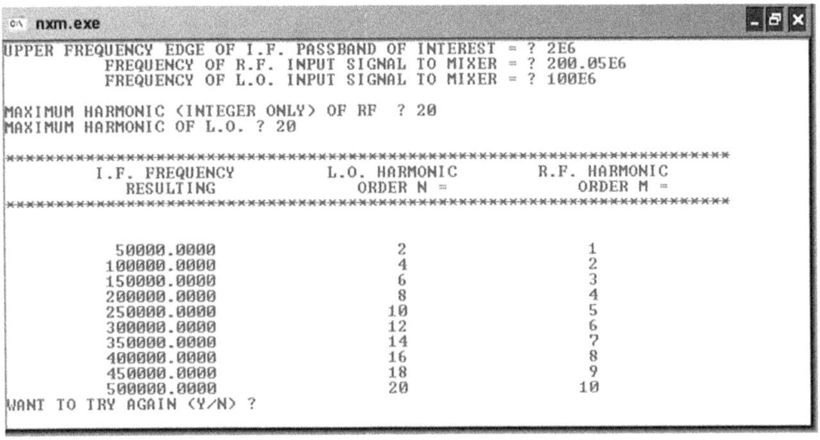

FIGURE 9.5 *N*-by-*M* cross products in a sampling phase detector output passband for a phase detector sampling freq = 100 MHz and a stray signal of 200.05 MHz (program = NXM.EXE in toolkit).

It is important to realize that a spur in the phase detector output that falls in the PLL passband can become highly magnified by the feedback division factor around the loop. In other words, a PLL using a 100 MHz phase detector and a VCO output frequency of 3 GHz will use a feedback counterfactor of 30; therefore, we will see cross-talking alias products magnified by 20 log 30 = 29.54 dB. This essentially means there is no margin for cross-talk in the system if the output spurious performance is to be −90 dBc. Who needs this kind of performance? Here is only one example of an application needing it: an FM broadcast transmitter. Suppose it is transmitting 100 kW out the antenna. Suppose its output is contaminated with a single spur 200 kHz away from its main carrier at −80 dBc. This spur (a pair of spurs actually at +/− 200 kHz away) would generate a spur power of 1 mW, not an insignificant amplitude. In fact, the FCC (Federal Communications Commission) has set the bar for performance to be on this order of magnitude. Many other applications that are sensitive to spurs include medical, radar, sonar, and communications receivers.

Applications highly forgiving to spurs may include clock recovery systems. For example, a DSB (double-sideband) incidental phase modulation of 0.1 radians (−26 dBc for each of the two mirror-image spurs) will yield only 16 ps of peak deterministic jitter at a clock rate of 1 GHz (calculated by multiplying {5.7/360} by the 1 ns period of a 1 GHz clock). So, for many high-frequency clock recovery applications one "can get away with murder." It would be hard for someone skilled in the art to produce a PLL that barbaric.

10

TIME AND PHASE JITTER, HETERODYNING, AND MULTIPLICATION

10.1 HETERODYNING AND RESULTING TIME JITTER

By heterodyning we mean the translation of a signal at frequency A to a new frequency B. The trigonometry is simple. The approach involves an RF mixer (akin to a multiplier) and a local oscillator. In some cases, two or more applications of heterodyning may be required to obtain the final result without spurs. Let us review the principles.

First, start with a single frequency conversion or "translation." Translation means the frequency slides from one value to another but all modulation features (if any) are preserved. To do this, we simply apply the unmodulated "shifting" signal to the L.O. port of an analog multiplier or RF mixer. Usually, this port is driven into clipping or saturation, so any AM on the L.O. will be stripped away anyhow. The reason why this port is overdriven is that the implementation of an RF passive mixer uses diodes that must be turned on and off by the local oscillator waveform. The other input to the RF mixer is the linear port, and thus any AM, FM, or PM is preserved. See Figure 10.1.

As one can see, multiplication (mixing) of one signal by another does not alter the modulation function, though it might suffer alteration in the bandpass filtering or subsequent processing.

So, let us assume some time jitter, say 10 ns (horrible) on a 10 MHz signal. This is equivalent to a phase modulation of 36° (0.628 rad). This phase modulation will remain unaltered after heterodyning. Now assume an L.O. frequency of 1 GHz and a mixer output desired sideband centered at (1000 − 10) MHz = 990 MHz, with the

Frequency Acquisition Techniques for Phase Locked Loops, Daniel B. Talbot.

Published 2012 by John Wiley & Sons, Inc.

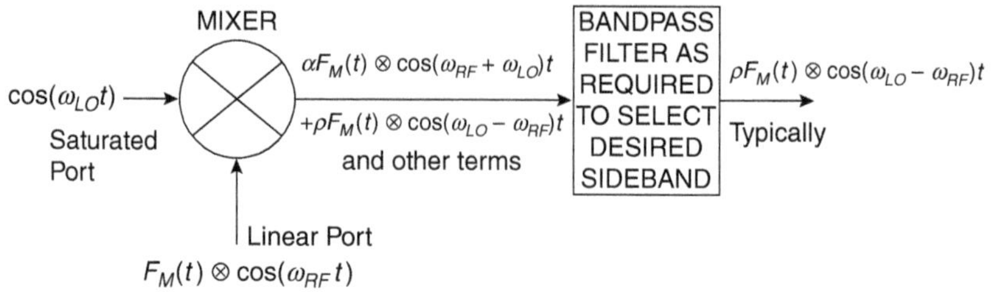

FIGURE 10.1 Heterodyne frequency translation preserves modulation function $F_M(t)$ at final output.

same PM sidebands as were present on the 10 MHz RF port. Since the 0.628 rad of PM is unchanged, the jitter has now shrunk by a factor of 99; it is now only 0.101 ns. *This is profound.* Essentially, it tells us that we can trade off jitter for phase noise if we frequency translate, and this fact has been exploited in a frequency-agile pulse generator for a client of this author.

In the example just cited, 10 MHz and its sideband ensemble was converted to 990 MHz having the same modulation sideband ensemble. This is called *upconversion*. If we had an RF signal (at the mixer's linear port) centered at 990 MHz, the output would have been centered at 10 MHz, and this would be called *downconversion* and the modulation sidebands would be reversed (upper becomes lower).

10.2 FREQUENCY MULTIPLICATION AND ANGLE MODULATION INDEX

If the modulated signal had been applied to a frequency multiplier, the phase or frequency modulation index β would increase by the multiplication factor. Since most frequency multipliers are nonlinear (saturated), AM would likely be destroyed.

A possible architecture for a frequency multiplier is a class-C amplifier. The output of such an amplifier is rich in harmonics and a bandpass filter or tuned circuit in the amplifier's output can select the desired multiplied center frequency. Assuming the bandpass filter just mentioned does not alter the multiplied phase or frequency modulation, we have a device that performs multiplication of both center frequency and PM and FM rather than mere translation.

If one cares about phase noise, frequency multiplication deepens that concern since the phase noise is elevated by 20 log N, where N is the multiplication factor.

10.3 FREQUENCY MULTIPLICATION'S ROLE IN CARRIER RECOVERY

A frequency multiplier is highly attractive as a device for recovering a carrier from a suppressed carrier digital modulation signal. For example, in biphase modulation, there are two statistically equal likely (assuming random data) phase states: 0

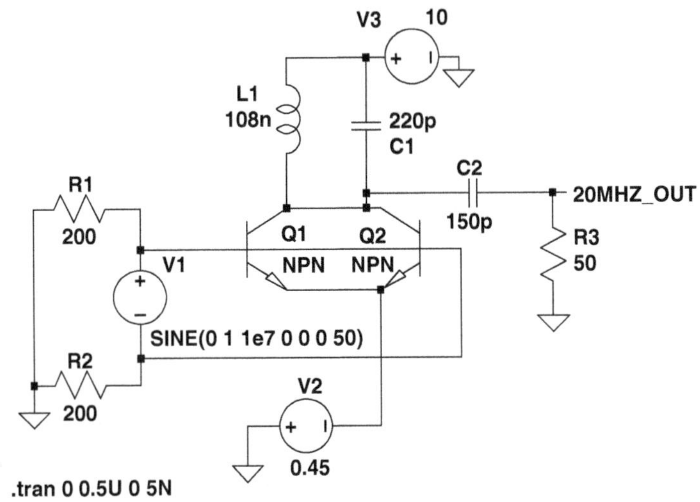

FIGURE 10.2 Push–push frequency doubler (10 MHz in; 20 MHz out).

and 180°. Since the carrier spends half its time (on average) in either state, the average carrier is zero. It is suppressed if observed over a long interval (or with small observation bandwidth). If observed using wide bandwidth (same as saying short interval), it is not suppressed. But if we apply this BPSK (biphase shift-keyed) signal to a frequency doubler, the zero-phase state is doubled to zero, but the 180° state is doubled to 360°. Thus, both phase states are now indistinguishable and there is now

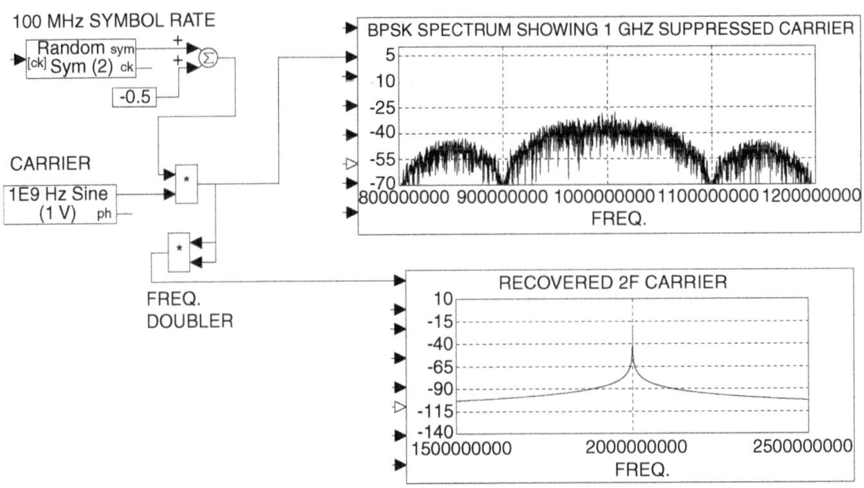

FIGURE 10.3 BPSK carrier recovery *via* frequency squaring (not shown is divide by 2 to obtain original 1 GHz).

a distinctly nonsuppressed carrier at twice the frequency. This double frequency carrier can then be applied to a divide-by-2 flip-flop and voila, we have generated a nonsuppressed carrier at our original invisible carrier frequency! This is called a frequency-squaring approach to carrier regeneration for a BPSK signal. Of course, it is a misnomer, because the frequency is doubled then halved, not squared at any stage in the architecture. What is squared is the waveform, since a class-C multiplier (or even a mixer with both inputs tied together, or a full-wave rectifier, called a push–push doubler) has a second-order exponent in its polynomial (Taylor series) expansion. The same concept can be applied to a quadriphase suppressed carrier, but this time a frequency multiplier factor N of 4 must be used. Since each phase jumps on average 90° at an equal likelihood, that is, 0, 90, 180, and 270, multiplication by 4 creates phase jumps of 0, 360, 720, and 1080°, respectively, all divisible by 360. This yields a constant phase carrier at four times frequency, but now all phases overlap, preventing carrier suppression. Thus, we can divide the result by 4 and recover our original carrier that was suppressed. Frequency multiplication also has a myriad of other uses. It is often used in frequency synthesizers, for example. Notice that a PLL having a divide by 2 or 4 in its VCO feedback path will double or quadruple a nonsuppressed carrier input but not a suppressed carrier input. For the latter, we need a class-C or other nonlinear device, active or passive (such as diodes).

Figure 10.2 illustrates a simple push–push frequency doubler, modeled in LTspice, file named “FREQ_DBLR.asc.”

Figure 10.3 shows a BPSK frequency-squaring system and the recovered carrier. The simulator is VisSim, and the file name is “BPSK_FREQ_SQ.vsm.”

11

CARRIER RECOVERY APPLICATIONS AND ACQUISITION

11.1 FREQUENCY MULTIPLIER CARRIER RECOVERY IN GENERAL

In the examples of the frequency multiplication approaches to recovering a suppressed carrier, the benefits were obvious. However, practical signals become degraded by filters and by noisy channels. Care is advised in selecting the frequency multiplier approach versus another approach called a Costas PLL.

Before we examine the Costas PLL, let's see how far we can "stretch" the application of frequency multiplication to carrier recovery. Let's try using X4 multiplication on some modulation formats. Start with 256QAM, then 16QAM, and then 4QAM and SQPSK (offset QPSK). See Figures 11.1–11.5.

As we've just seen, a X4 frequency multiplier can recover the carrier from any I and Q symmetrical data constellation. For the QPSK and OQPSK (SQPSK) cases, the carrier recovery is robust, as can be appreciated from the lower plot in Figures 11.4 and 11.5. In higher order modulation formats, the carrier recovery degrades in quality unless followed by a very narrow band PLL. If the PLL bandwidth needs to be extremely low, the possibility of tracking low-frequency transmitter phase twitching decreases. It is important to remember that even though the modulation format represents deliberate phase modulation, there is also parasitic PM in any signal source, and as Leeson's model shows, this PM is largest in spectral density at low frequencies. There is no such thing as a perfectly periodic signal

Frequency Acquisition Techniques for Phase Locked Loops, Daniel B. Talbot.

Published 2012 by John Wiley & Sons, Inc.

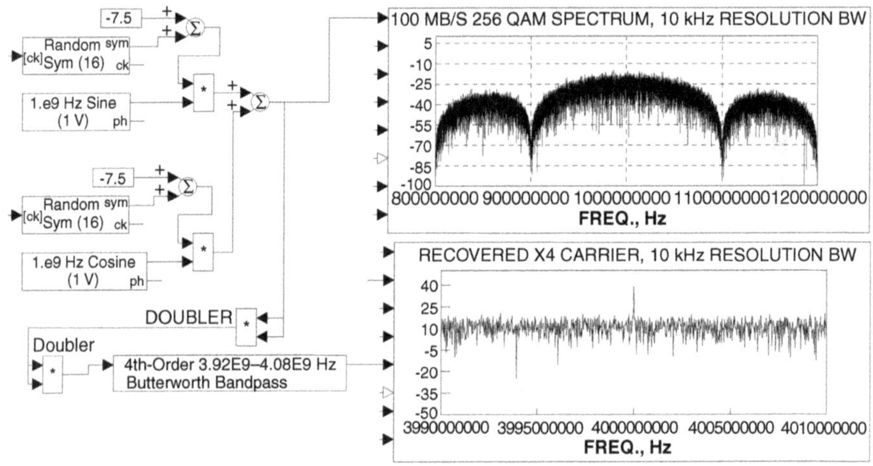

FIGURE 11.1 Carrier recovery from 256QAM using X4 frequency multiplication.

(carrier) except in textbooks; every signal source twitches slightly in phase and therefore frequency. Because of this, system engineers must take care in designing the coding. It is preferable to eliminate as much low-frequency energy from the code as possible, so that the carrier recovery system is not contaminated with code energy close to carrier. Frequency-multiplier carrier recovery is almost always followed by a PLL or in some cases by a narrow bandpass filter to improve SNR.

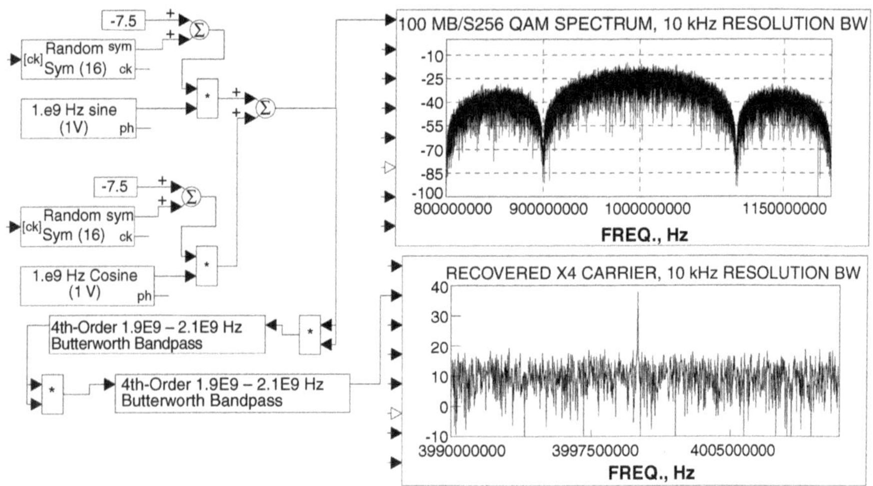

FIGURE 11.2 Carrier recovery from 256QAM using X4 frequency multiplication and interstage filtering between cascaded doublers.

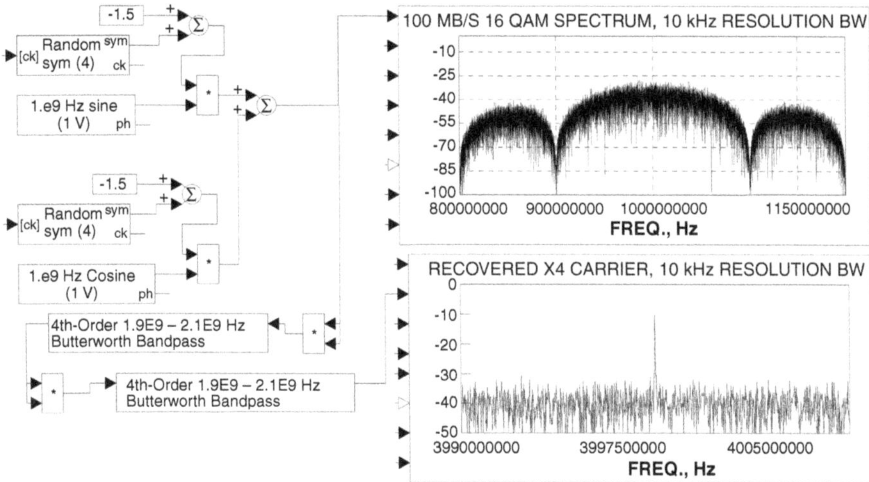

FIGURE 11.3 Carrier recovery from 16QAM using X4 frequency multiplication and interstage filtering.

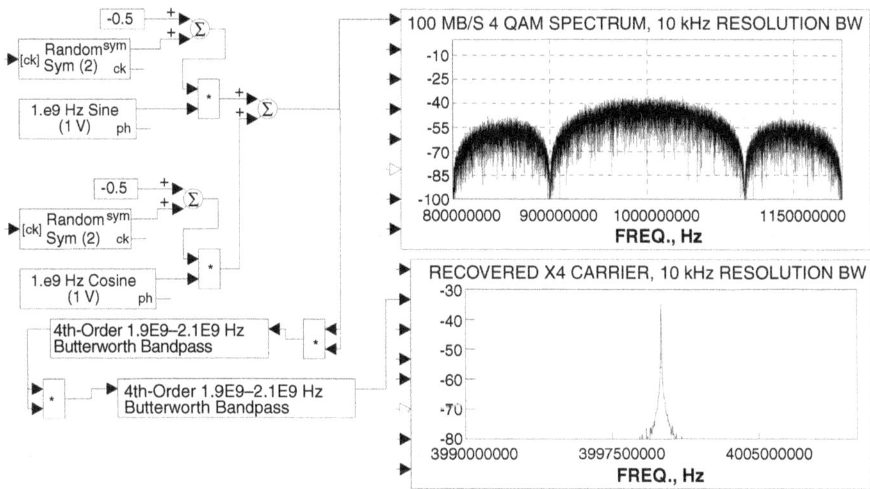

FIGURE 11.4 Carrier recovery from 4QAM using X4 frequency multiplication and interstage filtering.

11.2 THE SIMPLEST FORM OF COSTAS PLL

The simplest form of Costas PLL [20] is the one for demodulating a BPSK or DSBSC (double-sideband suppressed carrier) waveform. The advantage of the Costas PLL is that it incorporates implicit carrier recovery and filtering. The disadvantage is that the carrier recovery characteristic is not separable from the rest of the circuitry, thus

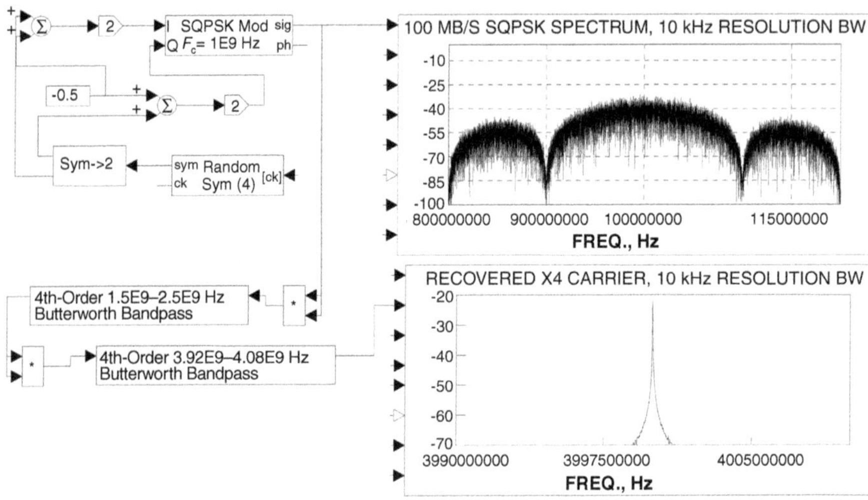

FIGURE 11.5 Carrier recovery from OQSPK (SPQSK) using X4 frequency multiplication and interstage filtering.

posing a slightly increased difficulty in implementing capture (carrier frequency acquisition) from a large VCO frequency offset that can be accomplished if a carrier recovery frequency-multiplier and sweeping PLL (or separate quadricorrelator) is used. We will discuss the quadricorrelator in more detail later. For carrier recovery SNR, it can be shown mathematically [21] that the SNR for a frequency-multiply-filter-divide approach is equal to the SNR of the Costas PLL approach assuming equal inertia (equivalent filtering).

We will next explain the DSBSC/BPSK Costas PLL. Refer to Figure 11.6. Let us assume for simplicity that the input $y(t)$ is a DSBSC signal, which can be written as

$$y(t) = \{\cos(\omega_c t)\} f_m(t) \tag{11.1}$$

where ω_c is carrier frequency and $f_m(t)$is modulation as a function of time having average DC=0.

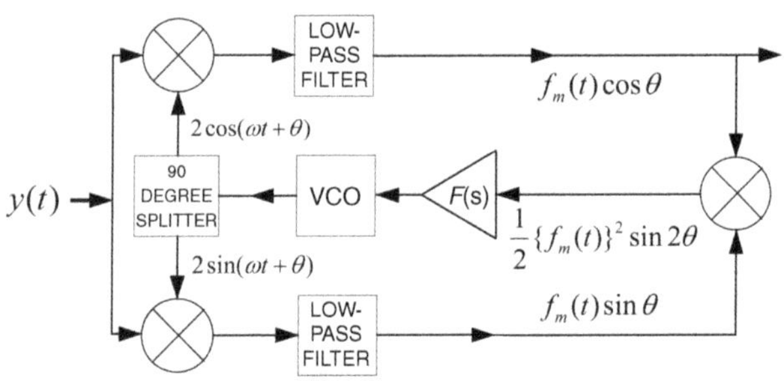

FIGURE 11.6 Simple DSBSC demodulation-type Costas PLL.

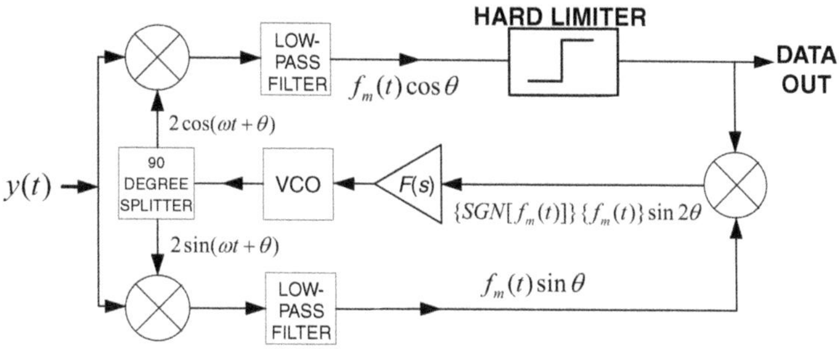

FIGURE 11.7 Simple BPSK demodulation slicing-type Costas PLL.

The carrier is implicitly recovered, and the modulation function is output as shown. For BPSK, there needs to be a comparator to slice the data (decide a one from a zero) anyway, so it might as well move inside the Costas PLL, which then resembles Figure 11.7; there is also a SNR benefit to incorporating the limiter. Thus, the Costas PLL serves dual roles, that of implicit carrier recovery and that of demodulation. Notice that the phase detector now outputs a voltage proportional to the sin of twice the recovered carrier error angle (2θ). Students of complex frequency and complex modulation will appreciate the equivalence of the Costas demodulator and the frequency multiply filter divide then demodulate approach. But whether frequency-multiply-filter-divide carrier recovery is used or a Costas loop is used, there will be a 180° phase ambiguity in the recovery for BPSK, and N times 90° ambiguity for QPSK.

11.3 HIGHER LEVEL QUADRATURE DEMODULATION COSTAS PLL

Redefining $y(t)$ to be a waveform describing QPSK or SQPSK (OQPSK) or some other orthogonal carrier signal and slightly increasing the circuit complexity yields the quadriphase Costas PLL of Figure 11.8. “Data Out 1” and “Data Out 2” are the demodulated waveforms.

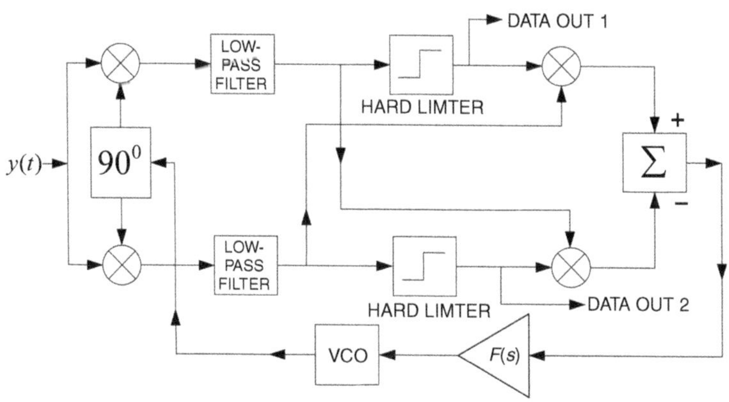

FIGURE 11.8 Quadriphase Costas PLL for demodulating QPSK.

11.4 FALSE LOCK IN BPSK COSTAS PLL

The portion of a BPSK Costas PLL excluding the closed loop is shown in Figure 11.9. We have deliberately avoided closing the loop so that we can inspect the tendency of the Costas PLL to falsely lock on a frequency-doubled data sideband. For this example, let us assume a data pattern that alternates between a 1 and a -1. Furthermore, assume a square wave pattern. If we assume the data rate to be 100 megasymbols/second, the highest frequency energy would occur for a 50 MHz square wave (each half of a 50 MHz square wave is one symbol). Thus, one would expect to see a pair of sidebands $+/-$ 50 MHz straddling the suppressed carrier. If we force the assumption that our VCO is 50 MHz off frequency, then the other sideband is now 100 MHz away (remember, the carrier is suppressed), generating a 100 MHz whistle in the phase detector output. This is also true for a non-Costas PLL. Such a PLL would be perfectly content to lock on a sideband 50 MHz above or below the suppressed carrier, and would contain a 100 MHz whistle in its phase detector output. Analysis has shown, however, that for the DSBSC Costas PLL, the same is true even for NRZ pseudo-random data, as shown in Figure 11.9. Notice the strong 100 MHz whistle dominating the phase detector output spectrum. Instead of forcing equilibrium at exactly 50 MHz offset, we have added an additional 200 kHz small offset

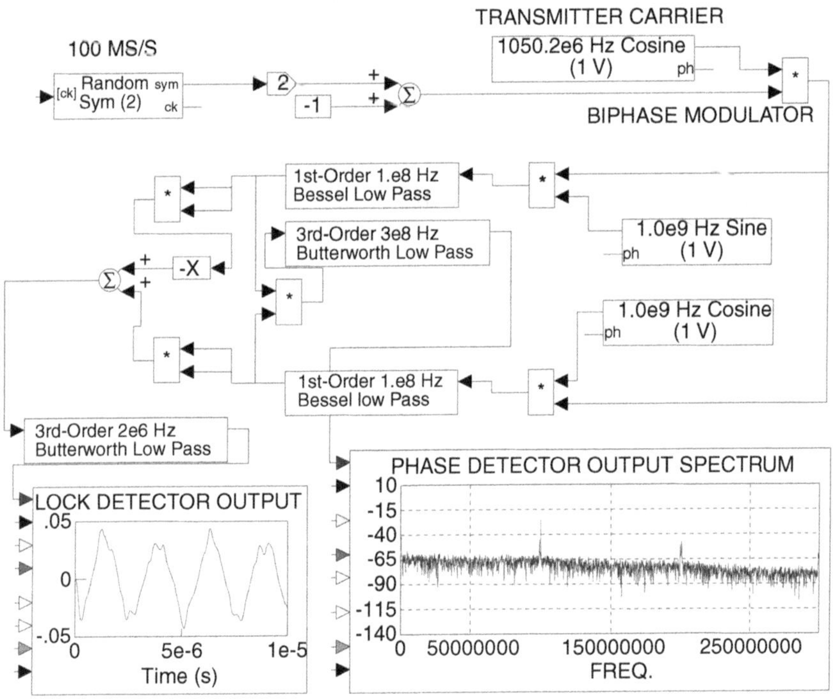

FIGURE. 11.9 Tendency for Costas PLL BPSK case to lock on a data sideband of $1/(2T)$.

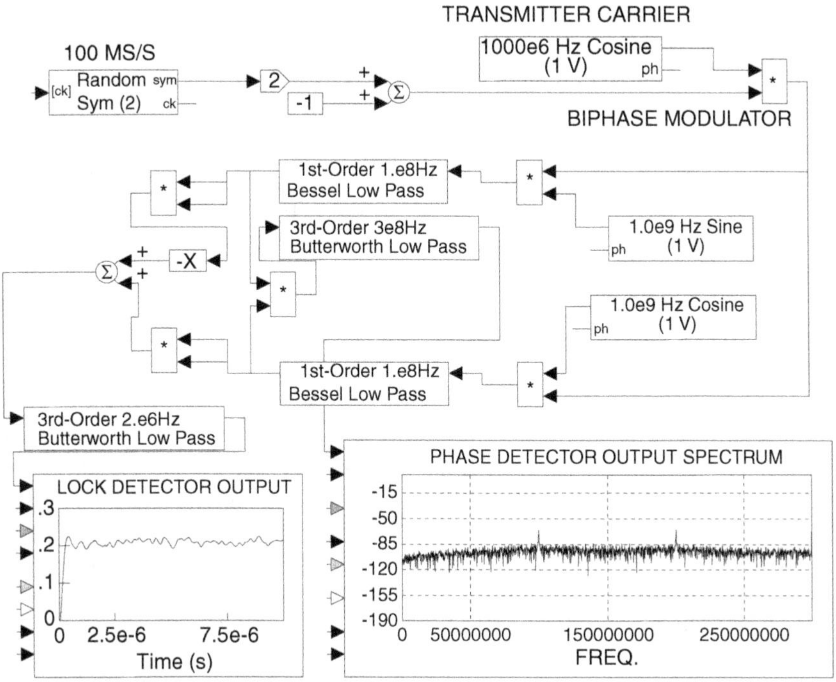

FIGURE 11.10 Costas PLL BPSK case not false locked on data sideband.

error so that we can watch the phase angle walk through two cycles for every cycle of 200 kHz (a characteristic of the DSBSC Costas PLL). Note that the lock detector for this case outputs a "lock indication" at each 180° of 200 kHz. Thus, the lock indicator seems to be fooled into producing a valid output at the 50 MHz offset false locking condition. However, this false lock indication voltage is not robust for the case we have simulated. Specifically, compare the amplitude of the false lock indication signal with the amplitude of the true lock indication signal (shown being 200 mV peak in Fig. 11.10). Notice also in Figure 11.10 that the amplitude of 100 MHz whistle in the phase detector output is almost insignificant by comparison to the amplitude of this signal under false lock.

It should be obvious that several assumptions were made to illustrate these examples. One assumption is that a fairly low-frequency low-pass filter is employed at the lock indicator circuit output. Another assumption is that AGC is accomplished elsewhere than by sensing the output of the "I" (in-phase) channel (the one to the left of the block that multiplies the DSBSC signal by a cosine of carrier). If AGC were accomplished via such coherent sensing (appropriately called coherent AGC), the false lock signal would become just as enthusiastic in amplitude as the true lock signal, fooling the false lock decision threshold and making the distinction impossible. Also, we used data filters having excess bandwidth. This is reasonable. In fact, the data filters may be external to the Costas PLL to improve false lock immunity. These differences may be partially responsible for the diverging conclusions of Olson versus Holmes, Hedin, Lindsey, and Woo [22], as well as filter relative bandwidths prior to slicing in the I and Q (quadrature) arms. *Note*: a first-order Bessel filter is simply a typically first-order RC (resistance–capacitance) filter. It is the same as any other first-order filter (Butterworth, Tchebyshev, Legendre, etc.). A more elegant filter is necessary in order to remove carrier harmonics and out-of-band signals more adequately. Such an increased filter order may alter results in a way that is probably best simulated rather than analyzed in closed form due to the increasing complexity (and opportunity for mistakes) in writing the time domain solution.

Another real-world condition is that frequency acquisition may be realized by sweeping the VCO frequency (in search of lock). Said sweep's speed may be too fast to allow significant energy build-up in the lock detector filter for the lower amplitude false lock indication. In fact, sweep speed may be used to advantage by presenting excessive acceleration to the PLL so that lock does not occur on nonrobust error signals, be they phase detector or lock indicator developed, but slow enough sweep speed to allow lock on robust signals. Hence, there are several degrees of freedom available to the designer.

Figure 11.11 shows the false lock amplitude and phase detector sideband whistle amplitude for the case where the "I" arm is hard limited. Amplitudes are comparable to the nonlimited condition. For those interested in a rigorous mathematical treatment of false lock, see Ref. [21].

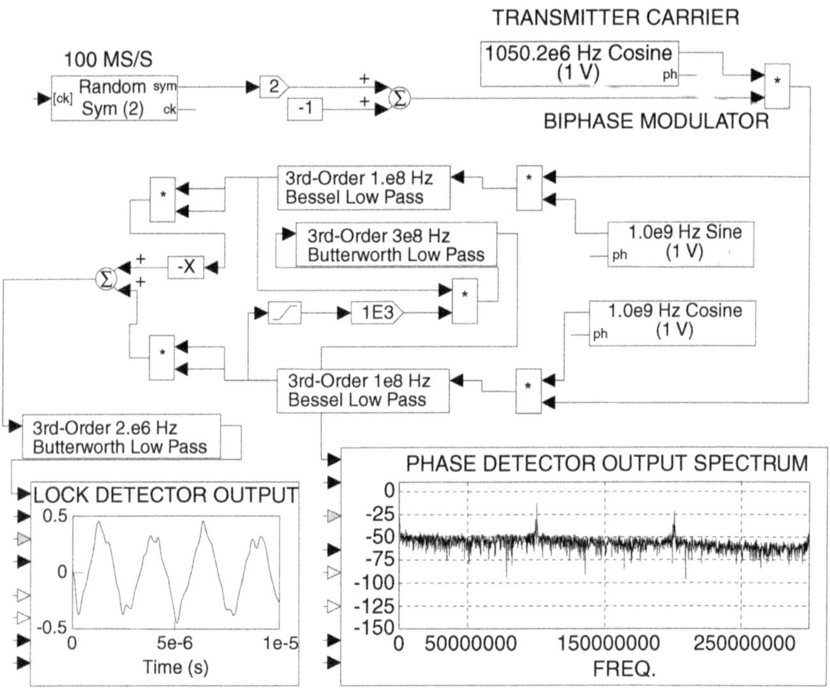

FIGURE 11.11 Hard-limited Costas PLL BPSK case potentially false locked.

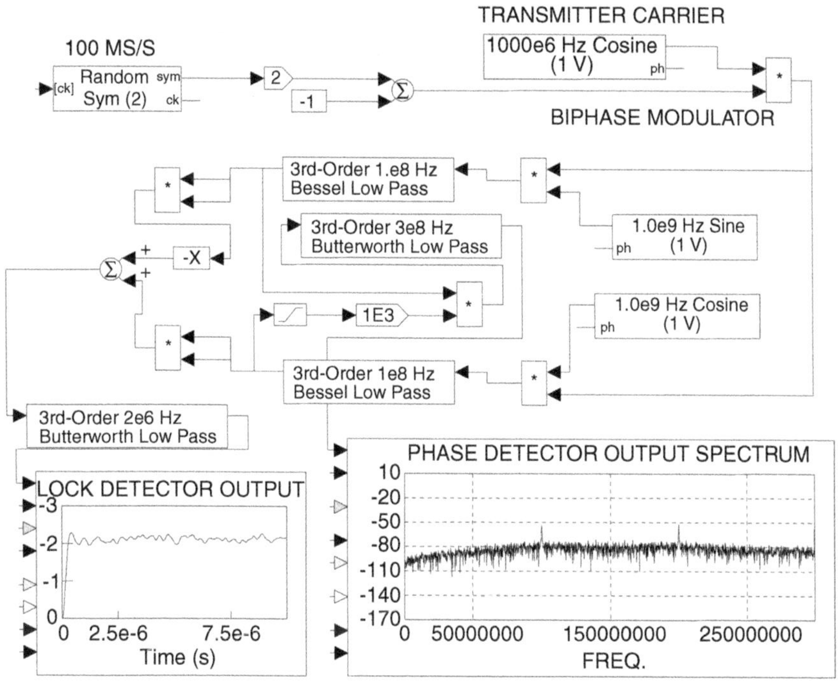

FIGURE 11.12 Hard-limited Costas PLL BPSK case not false locked.

Figure 11.12 shows a BPSK Costas PLL using hard limiting in the I channel and showing relative signal amplitudes from the lock detector and bit rate energy from the phase detector.

Figure 11.13–11.17 show a swept frequency acquisition Costas PLL with anti-false-lock circuitry responding to several types of code, all at 10 ns bit time. Figure 11.18 shows the same PLL using a code having a 5 ns bit time. One can see the sweep waveform prior to its extinguishing via the sweep gate as soon as true lock is verified. Also visible is the true lock signal, the phase detector whistle component at 100 MHz when the VCO frequency is off by 50 MHz, and then in Figure 11.19 we see the same swept Costas PLL but with the whistle frequency expectation filter set for 10 MHz, and a code having 100 ns bit time. We also see the whistle frequency of 10 MHz occurring at 5 MHz VCO offset, which occurs late in a sweep that begins at −55 MHz. Also shown is the whistle detection decision circuit output by itself (brown trace on right-hand plot).

11.5 ADDITIONAL MEASURES FOR PREVENTION OF FALSE LOCKING

Several beneficial measures may be helpful to prevent false locking. First, if the transmitted signal is “constant envelope,” that is, a signal whose modulation peaks

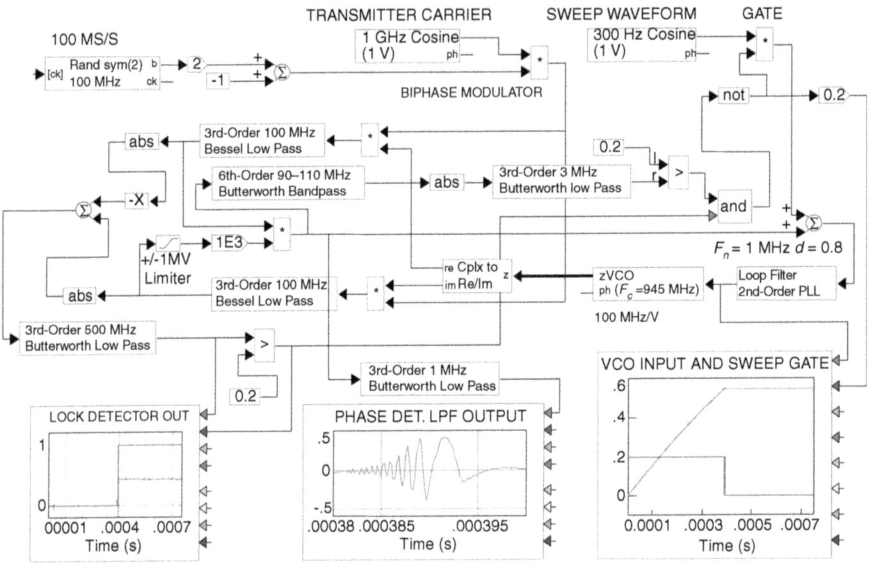

FIGURE 11.13 VisSim COMM 8 simulation of BPSK costas PLL with sweep, NRZ data and false lock prevention, coherent lock Detection. (*See the color version of this figure in color plates section.*)

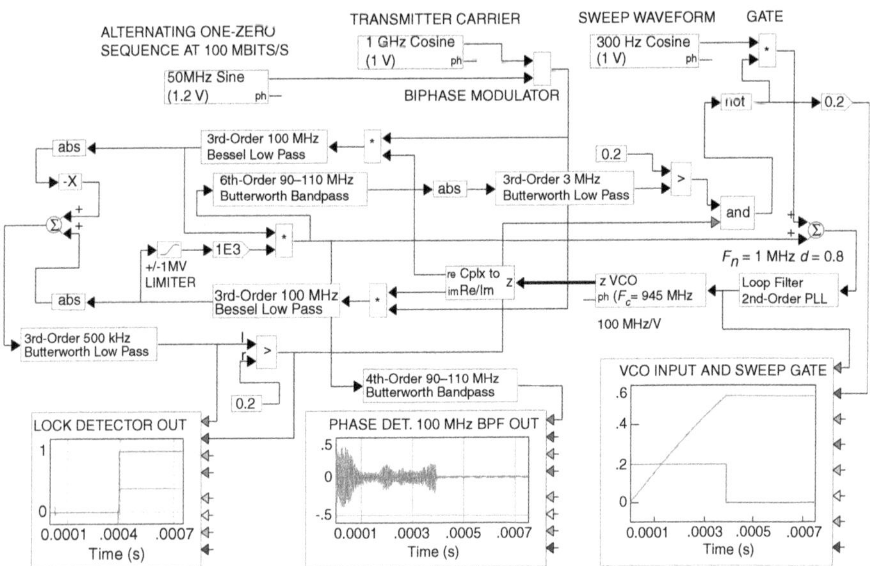

FIGURE 11.14 BPSK Costas PLL with false lock prevention responding to a 50 MHz square wave (equivalent to 100 Mbits/sec) modulation (note strong 100 MHz from phase detector but not from lock detector as VCO sweep crosses 50 MHz offset Frequency). (*See the color version of this figure in color plates section.*)

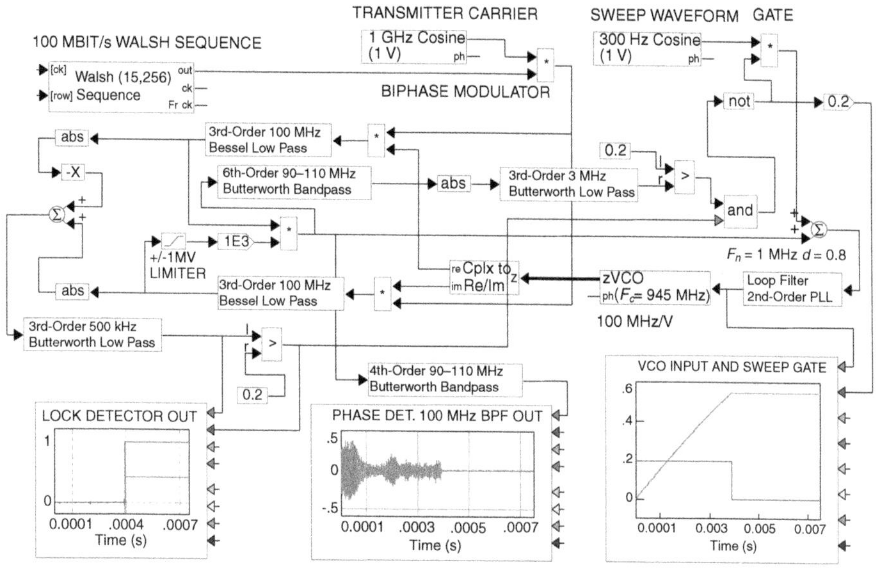

FIGURE 11.15 BPSK Costas PLL with false lock prevention responding to a walsh sequence (100 Mbits/s) modulation (note strong 100 MHz from phase detector but not from lock detector as sweep crosses 50 MHz offset frequency). (*See the color version of this figure in color plates section.*)

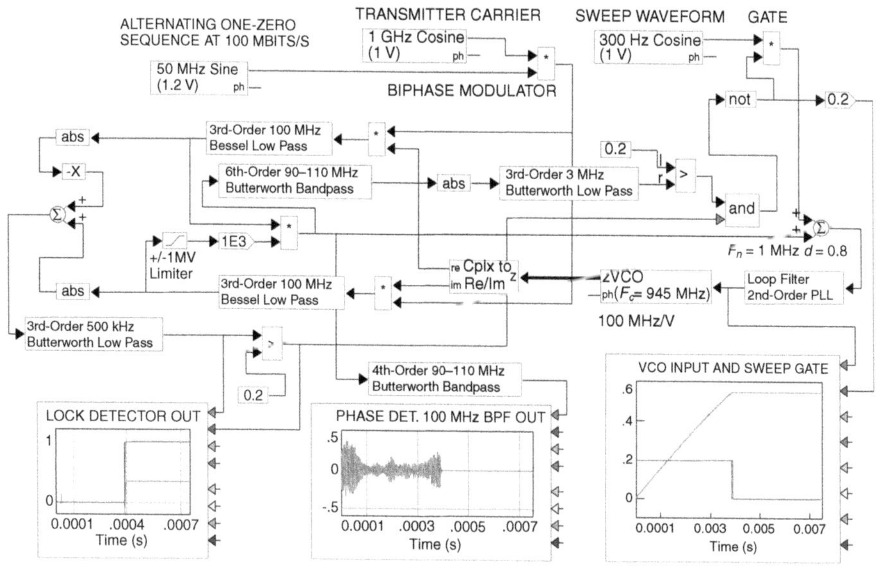

FIGURE 11.16 BPSK Costas PLL with false lock prevention responding to a 50 MHz (100 Mbits/s) sine wave modulation (note strong 100 MHz from phase detector but not from lock detector a sweep crosses 50 MHz offset frequency). (*See the color version of this figure in color plates section.*)

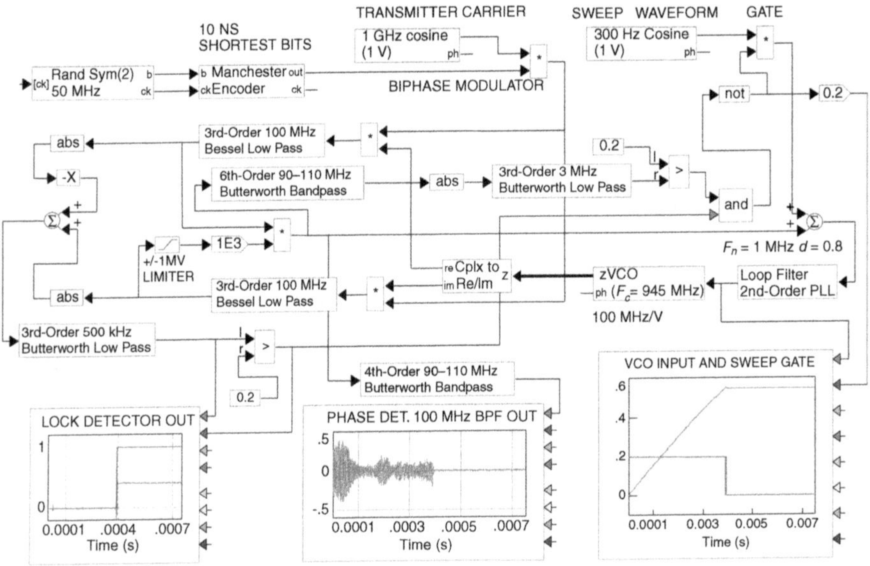

FIGURE 11.17 BPSK Costas PLL with false lock prevention responding to a 100 Mbits/s Manchester modulation (note strong 100 MHz from phase detector as sweep crosses 50 MHz offset frequency). (*See the color version of this figure in color plates section.*)

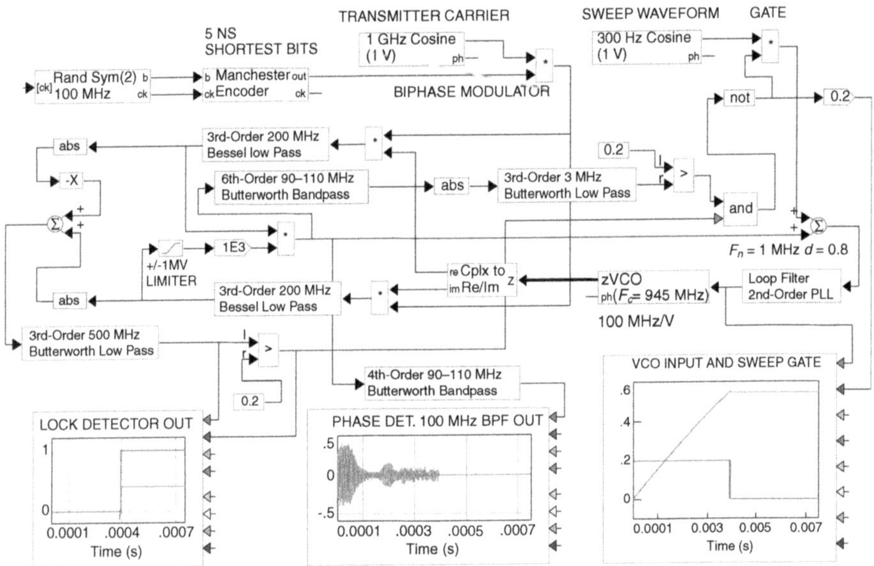

FIGURE 11.18 BPSK Costas PLL with false lock prevention responding to a 200 Mbits/s Manchester modulation (note strong 100 MHz from phase detector but not from lock detector as sweep crosses 50 MHz offset frequency). (*See the color version of this figure in color plates section.*)

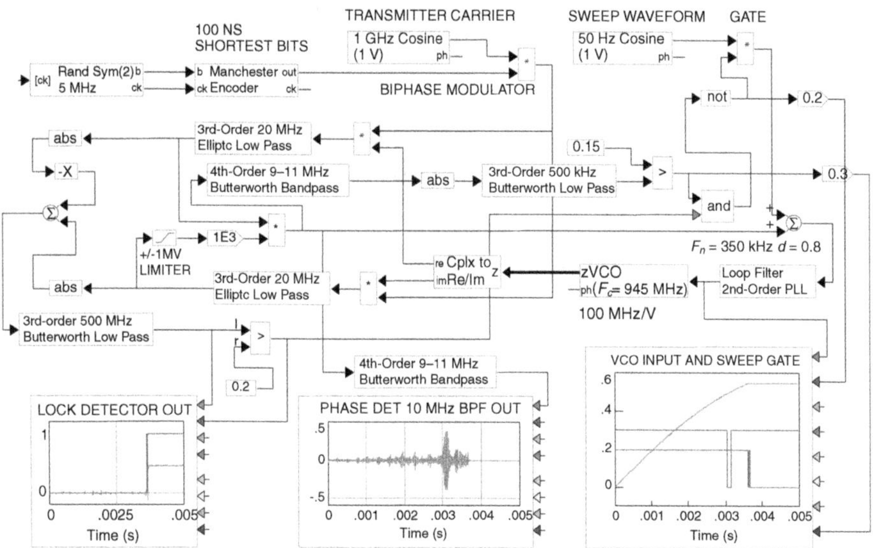

FIGURE 11.19 BPSK Costas PLL with false lock prevention and 10 Mbit/s Manchester modulation (note strong 10 MHz from phase detector but not from lock detector as sweep crosses 5 MHz offset frequency). (*See the color version of this figure in color plates section.*)

always have the same absolute value, the AGC loop can be replaced with a hard limiter at RF or IF (intermediate frequency). The hard limiter's advantage is its instantaneous action, and its simplicity. Of course, it must be preceded by the correct amount of bandpass filtering. Hard limiting avoids the exaggerating effects of coherent AGC derived from the "I" arm of the Costas loop when attempting to derive a lock indication. It also restores the "constant envelope" feature of the filtered data RF waveform, thereby reducing the data sidebands that cause false locking (compare Figs. 11.20 and 11.21 for a frequency-multiplier carrier recovery circuit).

Other measures include limiting the VCO sweep span so that it cannot lock onto a sideband (easily accomplished either by direct digital frequency synthesis, or DDS) or by using a VCXO (voltage-controlled crystal oscillator).

Another very good technique is to design a test sequence into the code, which when demodulated can authenticate whether true lock has been obtained. The result of the decision that data are valid can be used to stop the VCO sweep.

A frequency-squaring circuit may also prove a beneficial replacement for the Costas PLL, provided it is implemented open loop and *then* followed by a PLL to narrow-band its output spectrum and jitter. We have already shown that the frequency-squarer output is relatively devoid of data sidebands if preceded by minimal filtering. In fact, it has the advantage of operating on the transmitted signal spectrum, which is usually root-raised cosine rather than the raised cosine composite filtering (or some other additional filtering compared to the transmit signal) occurring in a Costas PLL. The Frequency-squaring circuit can also be

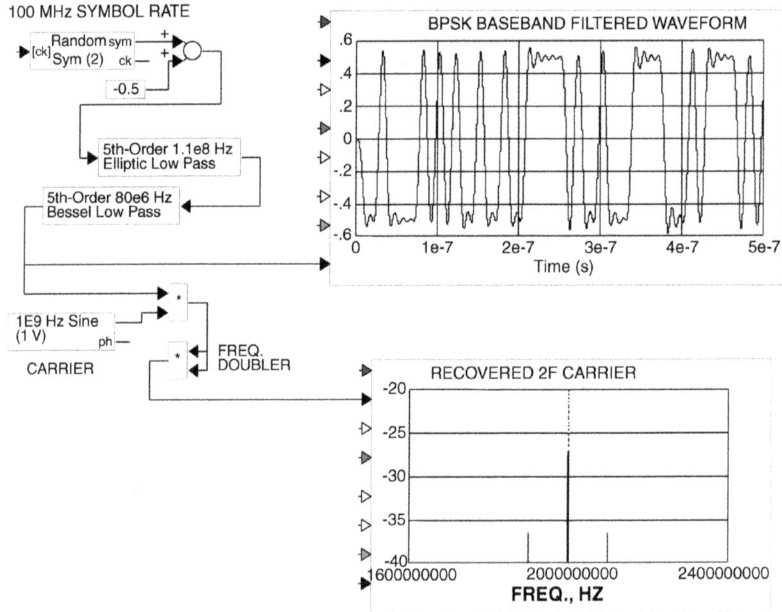

FIGURE 11.20 Bandpass filtered BPSK carrier recovery via frequency squaring (not shown is divide by 2 to obtain 1 GHz).

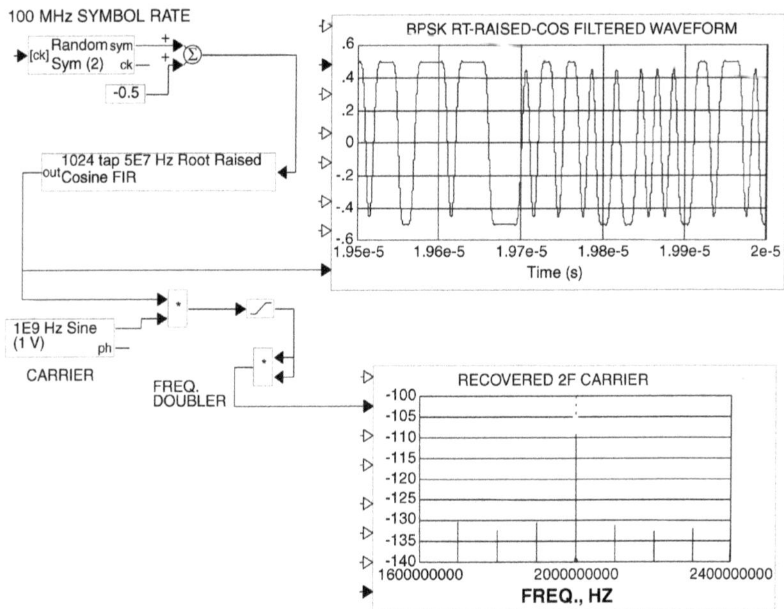

FIGURE 11.21 Root-raised cosine BPSK carrier recovery via frequency squaring with hard limiter before frequency multiplier (not shown is divide by 2 to get 1 GHz).

made to have reduced data-related sidebands by hard limiting as in Figure 11.21. An improvement in false locking for a subsequent PLL is thus achieved (compared to the case without the hard limiter, as in Fig. 11.22). However, the demodulator must preserve the root-raised cosine filtering from the transmitter and add its own in order to accomplish total raised cosine filtering for best ISI (intersymbol interference). Fortunately, the frequency-squarer doesn't have to concern itself with data recovery, only with carrier recovery. After the carrier second harmonic is recovered, it must be divided by 2, and in doing so, PM sidebands are reduced another 6 dB. (Caution: if hard limiting is employed ahead of a Costas PLL, the overall filtering transfer function may be affected. Phase response will be preserved but not amplitude response.)

Notice that undesirable data-related sidebands still exist even when hard limiting is used; this is due to the phase modulation of the transmitted signal, which, though unintentional, is a result of practical transmit filtering. Hard limiting restores constant envelope behavior but the incidental phase modulation (ICPM) remains. But since such a radical improvement (reduction) in sidebands is realized using hard limiting (about 7 dB improvement in the Figure 11.21 and 11.22 examples), we can more easily distinguish false lock. We will discuss false lock prevention using DC offset next.

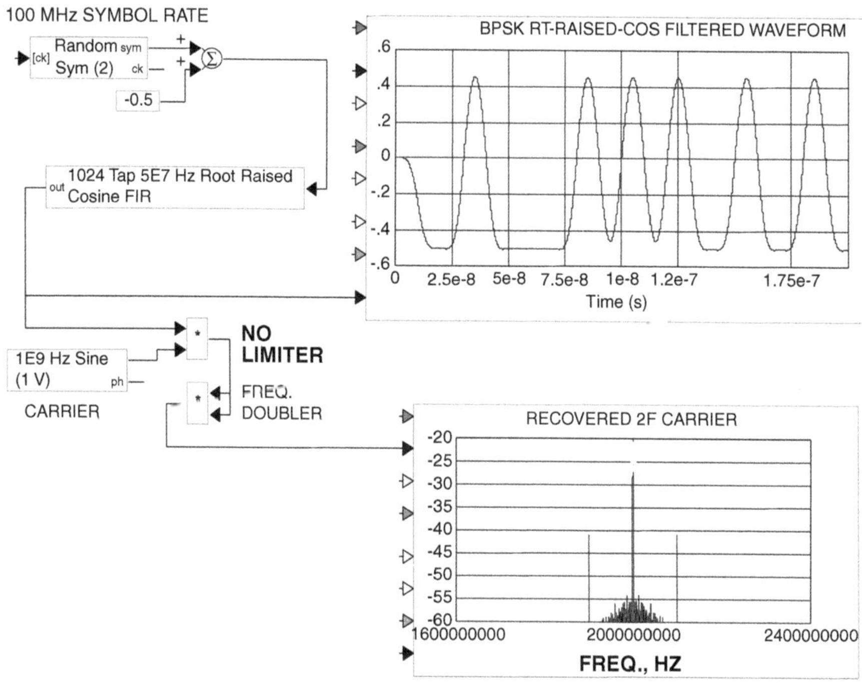

FIGURE 11.22 Root-raised cosine BPSK carrier recovery via frequency squaring with no limiter (not shown is divide by 2 to get 1 GHz).

11.6 FALSE LOCK PREVENTION USING DC OFFSET

Observe Figure 11.23, which shows a frequency-squarer feeding a conventional (not a Costas) PLL. Alternatively, we could divide the squarer output frequency by 2 prior to application to the PLL (using an octave lower VCO frequency) to gain a 6 dB advantage in carrier-to-sideband ratio before PLL processing. If the PLL is made type II, then the phase detector output DC value at lock is zero volts. If we introduce a small DC offset, the phase detector output will cancel that offset by producing an equal and opposite DC value. If this DC offset is sufficient in magnitude, false locking will be impossible because the sideband cannot generate adequate error signal amplitude from the phase detector output to negate said offset. Of course, the PLL will now lock only on the desired largest signal, but the drawback is that the phase angle of lock will have an offset. This offset can be negated and even adjusted by a passive phase shifter at RF. At very weak SNR, there will also be a noise bias in the PLL, possibly affecting acquisition and even hold-in performance. Generally, however, SNR will not be a problem. If noise bias is a concern, there are better techniques than the DC offset trick. A separate quadricorrelator is one alternative. Since the quadricorrelator is an FM detector, the response to AWGN (additive white Gaussian noise) near center frequency is essentially near zero (recall the FM bowtie effect discussed earlier). Thus, DC output from the ideal quadricorrelator after locking is zero with no DC offset due to noise.

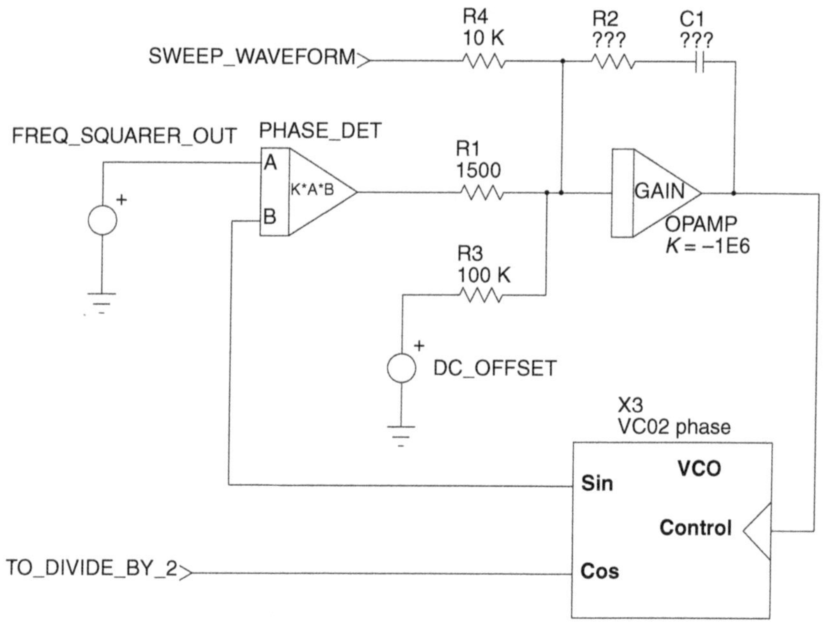

FIGURE 11.23 PLL with DC offset false lock prevention approach (shown operating at the 2F frequency squarer output).

12

NOTES ON SWEEP METHODS

There are a few caveats to observe when designing a sweep technique for a PLL. The discussion below applies to nonsuppressed carrier PLL sweep systems, and to Costas PLL systems with adjusted conditions and expectations.

12.1 SWEEP WAVEFORM SUPERIMPOSED DIRECTLY ON VCO INPUT

It is possible to superimpose a slow sweep waveform on the VCO tuning input without ever disconnecting it. Obviously, the amplitude and rate must be appropriate and compatible with acquisition (more on that later). In this method, it is assumed that the PLL will generate enough corrective force to cancel the frequency sweep (recall our discussion of a disturbance signal summed with the VCO input and the simulation via LTspiceIV). The cancellation of frequency error is perfect, but phase will not cancel perfectly. This will cause a wobbling phase at the sweep rate. This may be small or large, depending on sweep magnitude and speed. Furthermore, if the PLL is type II and unlocked, its op-amp-based loop integrator may be pegged against the power rail. In such a case, sweeping the VCO is too little too late; the loop integrator is stuck at the rail and therefore a correctly shaped beat waveform is impossible, since only a waveform can be developed whose voltage points away from the rail. It is generally required that the PLL be type I for that reason. Figure 12.1 illustrates direct VCO sweeping. If this type of sweep method is used, the next level of sophistication is the disconnect-when-locked approach.

Frequency Acquisition Techniques for Phase Locked Loops, Daniel B. Talbot.

Published 2012 by John Wiley & Sons, Inc.

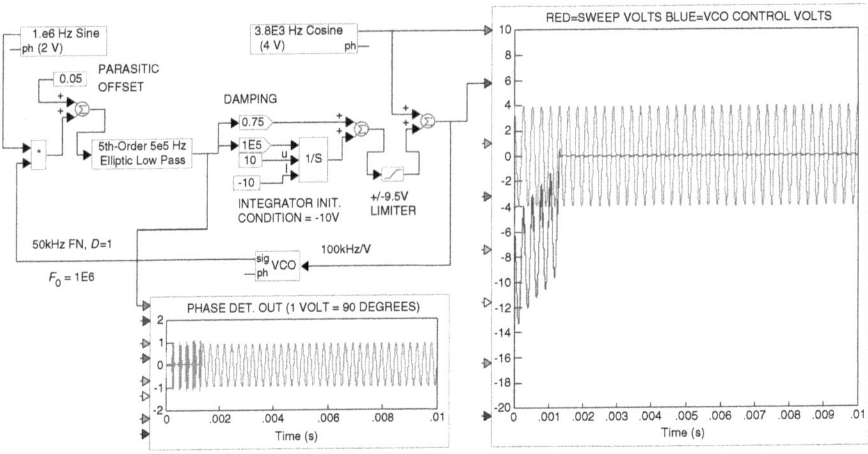

FIGURE 12.1 Maximum accelerating VCO sweep (note phase detector output shows peaks equal to 90° peak phase excursions). (*See the color version of this figure in color plates section.*)

Unfortunately, the problem is that when the sweep signal is disconnected there can be a discontinuity of voltage (a step, pop, or transient). This disturbs the loop, possibly to the degree of breaking lock after lock has been established for some point on the sweep waveform. This can be avoided by stair stepping the sweep signal, using a D/A (digital to analog) converter, and killing the update clock when lock has been achieved, so that the VCO baseline is now "parked" at the lock-friendly voltage level. Equivalently, a DDS (direct digital frequency synthesis) approach can be used. The advantage of a DDS as a VCO is that its output frequency is a well-known value determined by the digital code value applied to it. This code can be restricted and/or swept. Thus, a DDS output frequency can be managed via numerical codes to avoid points of false lock.

12.2 MAXIMUM SWEEP RATE (ACCELERATION)

We cannot sweep the PLL VCO frequency too fast, as previously mentioned, before it slips cycles. Most textbooks give an equation but forget to stipulate the assumptions. The most common assumption is that the phase detector used has 90° clipping (rollover). Another assumption is that the damping factor is between 0.707 and 1. Also, an implication is that the sweep maximum rate is a usable rate. Of course that is not true. The maximum sweep rate is a "drop dead" number. Because of DC offsets in either the phase detector or the loop integrator or both, or waveform distortion, repeated cycle slippings result in more than mere break-up or spectral splatter; these slippings may not average to zero volts, causing the loop integrator to drift quickly to the rail, further precluding graceful recovery when or if the sweep is disconnected.

The equation given for the maximum sweep rate for the assumptions outlined is

$$\frac{\Delta\omega}{\Delta t} = \dot{\omega} = \omega_n^2/2 \tag{12.1}$$

Let's assume a type-II PLL having damping of 0.707 and a 100 kHz loop natural frequency (not the same as 3 dB loop bandwidth). Thus, $\omega_n = 2\pi \cdot 100\,\text{kHz} = 628\,\text{krad/s}$, and the maximum permissible sweep rate for 100% probability of lock on one sweep is $\dot{\omega} = 1.97E + 11$ or 31.4 GHz per second. Thus, for a 1 GHZ VCO center frequency having a 10% uncertainty (100 MHz), we must sweep 100 MHz peak to peak in more than 318 μs. For a triangular sweep, this equals a sweep waveform whose frequency is 1570 Hz at a peak-to-peak amplitude corresponding to not greater than 100 MHz, or a 785 Hz triangle wave whose peak-to-peak amplitude is no more than 200 MHz. This rate is applicable only for acquisition. For tracking, the rate depends on modulation type and rate [1,23].

A PLL designed to demodulate an FM signal will fall out of lock (depending on SNR) or slip cycles when the modulation rate exceeds a limit, depending on damping, loop bandwidth, and phase detector rollover points, as well as clipping points anywhere in the loop. This normally precludes demodulation of FSK (frequency shift key) signals where the frequency is "popped" from one value to another. For such a case to succeed (and for the FM threshold improvement advantages of a PLL FM demodulator to be realized), the rising and falling edges of the baseband FSK modulation must be tapered to obey some acceleration limit. Unwillingness on the part of the modulator designer to do this will assure failure of the overall system (which has happened and the PLL consultant got the blame).

Figure 12.2 shows the maximum acceleration being slightly exceeded. Notice the effect of DC offset, resulting in asymmetric clipping and a "lock–unlock" dynamic

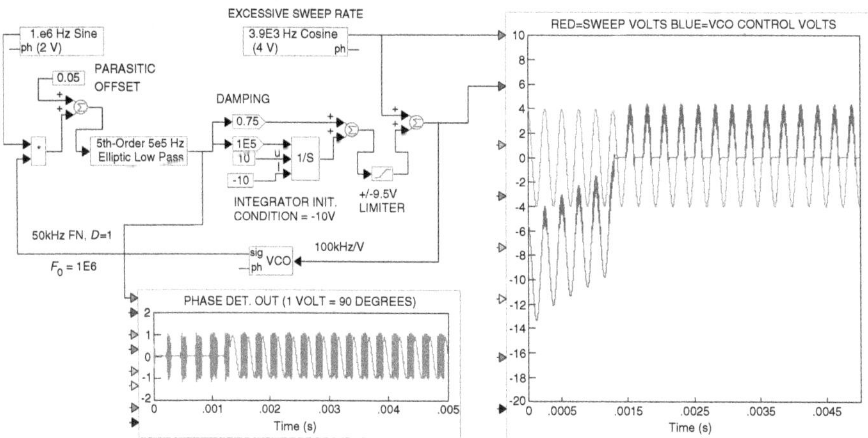

FIGURE 12.2 Excessive sweep rate (acceleration) causes phase detector rollovers (cycle slips) and splatter (positive peaks are expanded and negative peaks clipped due to phase detector's output voltage offset of 50 mV). (*See the color version of this figure in color plates section.*)

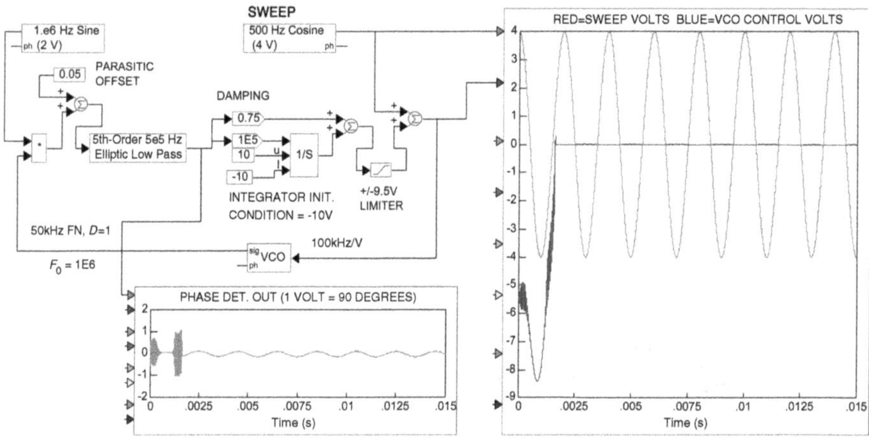

FIGURE 12.3 VCO sweep runs continuously; causes appreciable phase wobble of ~30° peak; also notice phase detector output offset to force integrator to drift in favorable direction (initially pegged at -10 V rail, damping pegged at -9.5 V); sweep rate = 500 Hz at 4 V peak = 1.257 GHz/s at sweep zero crossings. (*See the color version of this figure in color plates section.*)

during sweep. Also notice the severe cycle slips from the phase detector. This situation yields a PLL that may or may not lock depending on DC offsets.

Figure 12.3 shows a PLL with modest sweep rate and a 50 mV DC offset favorable for the starting phase chosen for the sweep (sweep is a cosine waveform). This offset would be unfavorable if the sweep had begun at its negative peak. The integrator output was set to be -10 V initially, as can be seen on the plot (VCO control voltage starts from a negative value offset by the 4 V positive peak of the sweep waveform). Lock is achieved during the first rising slope of the sweep waveform.

Figure 12.4 illustrates a PLL with inconsequential offset and a loop integrator initial condition of zero volts. Note how the VCO control voltage begins at approximately the sweep peak but converges to lock (at zero volts) rapidly after the first falling slope of the sweep waveform. A system having an integrator initial condition of zero volts is improbable if not done deliberately by an initializing circuit.

The sweep waveforms for Figures 12.3 and 12.4 are identical, as are the dynamic loop coefficients; thus, we would expect identical steady-state phase wobble (which is verified by the phase detector plot for each case). But by drastically reducing the sweep rate and amplitude as in Figure 12.5, we can get a phase wobble that is smaller by a factor approaching the square of the sweep frequency reduction factor (the loop is a 12 dB/octave high-pass filter) and the first-order reduction proportional to the sweep amplitude reduction. Therefore, for the example we have reduced sweep frequency by a factor of 2.5, and the square is 6.25; we have also reduced the amplitude by a factor of 4. Thus, we expect a phase wobble reduction of 25 times. Comparing the phase detector output plots of Figures 12.4 and 12.5 demonstrates that this ratio is correct.

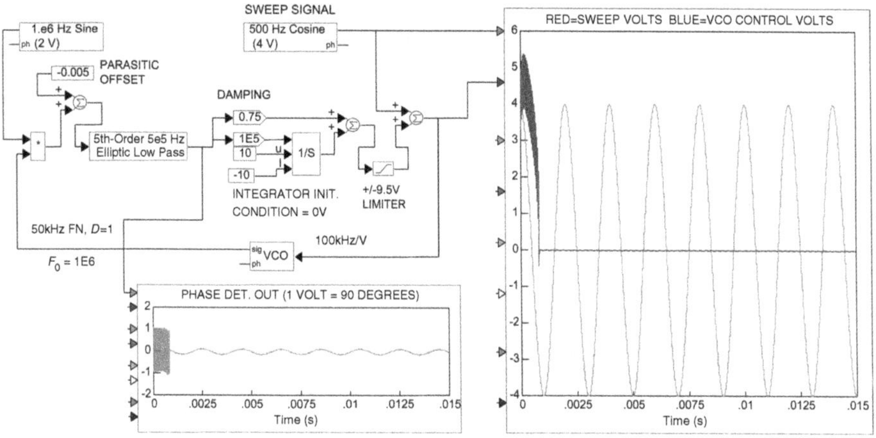

FIGURE 12.4 Integrator zero volts initial condition with -5 MV DC input offset, sweep signal continuously applied (no lock detector circuitry; note phase wobble of ~30° peak). (*See the color version of this figure in color plates section.*)

12.3 FALSE LOCK DUE TO HIGH-ORDER FILTERING

Observe the sorry situation of Figure 12.6. The average DC value of the VCO control voltage (shown in blue) seems to be stationary and satisfied, indicating that the sweep voltage is being canceled, that is, the loop is tracking and locked. But clearly it is locked at 500 kHz too high a frequency. What is causing this? It is caused by our use of wide sweep span combined with a very high-order sideband suppression filter. In

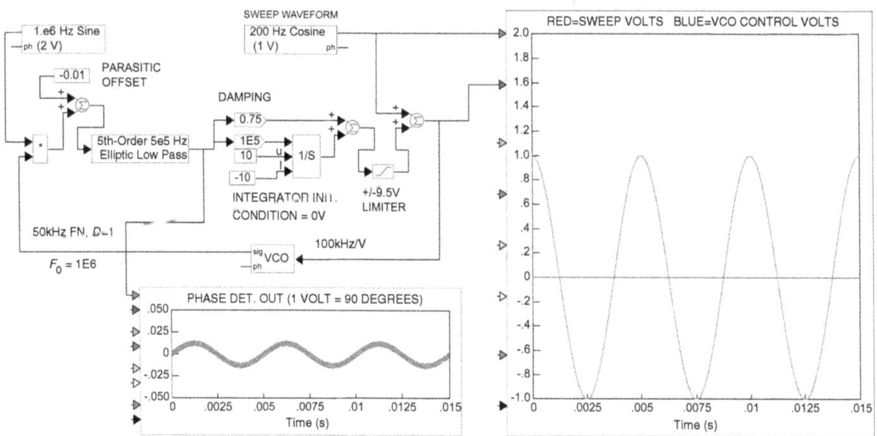

FIGURE 12.5 Comfortably slow sweep rate resulting in ~10° phase wobble when sweep runs continuously and not disconnected by a lock detection circuit (lucky acquisition, not recommended to combine type-II PLL with continuously running VCO sweeping). (*See the color version of this figure in color plates section.*)

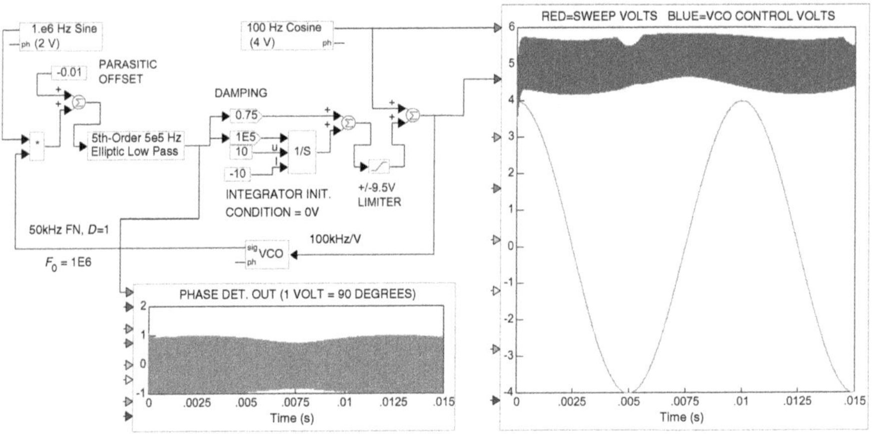

FIGURE 12.6 Unsuccessful phase lock in a type-II PLL with continuously running sweep due to false self-locking. (*See the color version of this figure in color plates section.*)

fact, the 270° lagging phase of the elliptic filter chosen occurs at about 500 kHz that corresponds to the frequency of false lock shown. We can experiment with the filter. Select the VisSim file titled "PLL-EXAMPLE-SELF-SWEEP-FALSELOCK-PREV .vsm" and run it using the VisSim viewer tool provided. Change the elliptic filter order to 3. Notice the false locking vanishes *for this sweep span*. Now change the filter order back to 5 but make the cutoff frequency 1 MHz rather than 500 kHz. Now, rerun the simulation and notice that the false locking behavior is gone and replaced by true locking at a VCO voltage of zero volts. Either we keep the elliptic filter as it is and limit the VCO control voltage to somewhere around +/− 1.5 volts or we reduce the phase lag of the filter by reducing its order and/or increasing its cutoff frequency. Always simulate for the worst case range of control voltage possible.

This false "self-locking" in a PLL occurs without any external spurious signals or sidebands. It can occur both due to delay within the loop and due to sideband suppression filtering (or an ensemble of parasitic poles) [23].

In the example shown, one might ask the obvious question "why use a sideband suppression filter at all?" That is a perfectly reasonable question. In the example, the drawback to not using a filter is simply elevated second and third harmonic at the VCO output, which can easily be removed if it is a fixed frequency. But if the PLL is frequency-agile or is a frequency multiplication PLL (one in which a divide-by-N counter exists in the VCO feedback path to the phase detector), removal of the sidebands *outside* the PLL becomes difficult if not impossible. Let the phase detector operating frequency be f_{REF} and the feedback divider factor be N. The output frequency is then N times f_{REF} and the sidebands are "potentially" present at

$$f_{\mathrm{SPURS}} = N \cdot f_{\mathrm{REF}} \pm M \cdot f_{\mathrm{REF}} \tag{12.2}$$

where M is any harmonic of the phase detector operating frequency (M can be a range of integer values, especially in the case of a sampling or edge-triggered phase detector). For example, suppose we need an output frequency of 5 GHz locked to a phase detector reference frequency of 100 MHz. This is satisfied by making $N = 50$. If we use a PFD (phase frequency detector), it will generate a comb spectrum of harmonics of 100 MHz. Thus, M ranges from an integer of value one to a very large integer. Furthermore, if N can take on integer values from 40 to 60 (creating a 4–6 GHz output frequency range in 100 MHz increments), the sidebands around the output frequency move with it, making it a fool's errand to attempt to remove them outside the PLL. Thus, in general it is necessary to incorporate a sideband suppression filter within the PLL.

12.4 SWEEP WAVEFORM APPLIED DIRECTLY TO PLL LOOP INTEGRATOR

In all examples prior to this chapter, we have shown sweep signals being applied directly to the PLL loop integrator input. This has numerous advantages. First, it assures that a type-II PLL can be employed, because the sweep waveform can, at some voltage and time instant, negate any "against the rail" loop integrator condition. Second, when it is disconnected by the sweep gate, there is minimal transient step or pop. It is the ideal sweep injection point. However, its appearance prior to the loop integrator slightly complicates the calculation of maximum acceleration in the PLL. The loop integrator time constant provides gain or attenuation to this sweep waveform.

Another advantage to preloop integrator sweep injection is that when it is disconnected, the static phase angle of lock is not wobbling while trying to "track out" the sweep, because disconnect can be accomplished by a simple analog switch driven from appropriate lock detection circuitry.

12.5 SELF-SWEEPING PLL

It is possible to deploy a PLL that implements its own sweep waveform by exploiting the behavior of the loop integrator. The schematic for a complete self-sweeping harmonic locking PLL is shown in Figure 12.7. A dual op-amp (U2A and U2B) serves as loop integrator (U2A) and as Schmitt trigger (U2B) fed back around the loop integrator, which causes the loop integrator to sweep back and forth limited by the Schmitt trigger input hysteresis window. A diode (D3) appears at the Schmitt output to assure more current injected into the loop integrator when the Schmitt circuit output is negative. This causes a high-speed rising edge sweep waveform compared to the falling edge, thus generating a sawtooth sweep waveform. The criteria for setting the sweep current injected into the loop integrator is simple. For the sawtooth slope on which locking is desired, the sweep current must be $\ll$ than the phase detector buffer's output current injected into the loop integrator. For the sawtooth slope on which locking is forbidden, the sweep current must be diode-steered so that it is $\gg$ than the phase detector buffer's output current injected into the

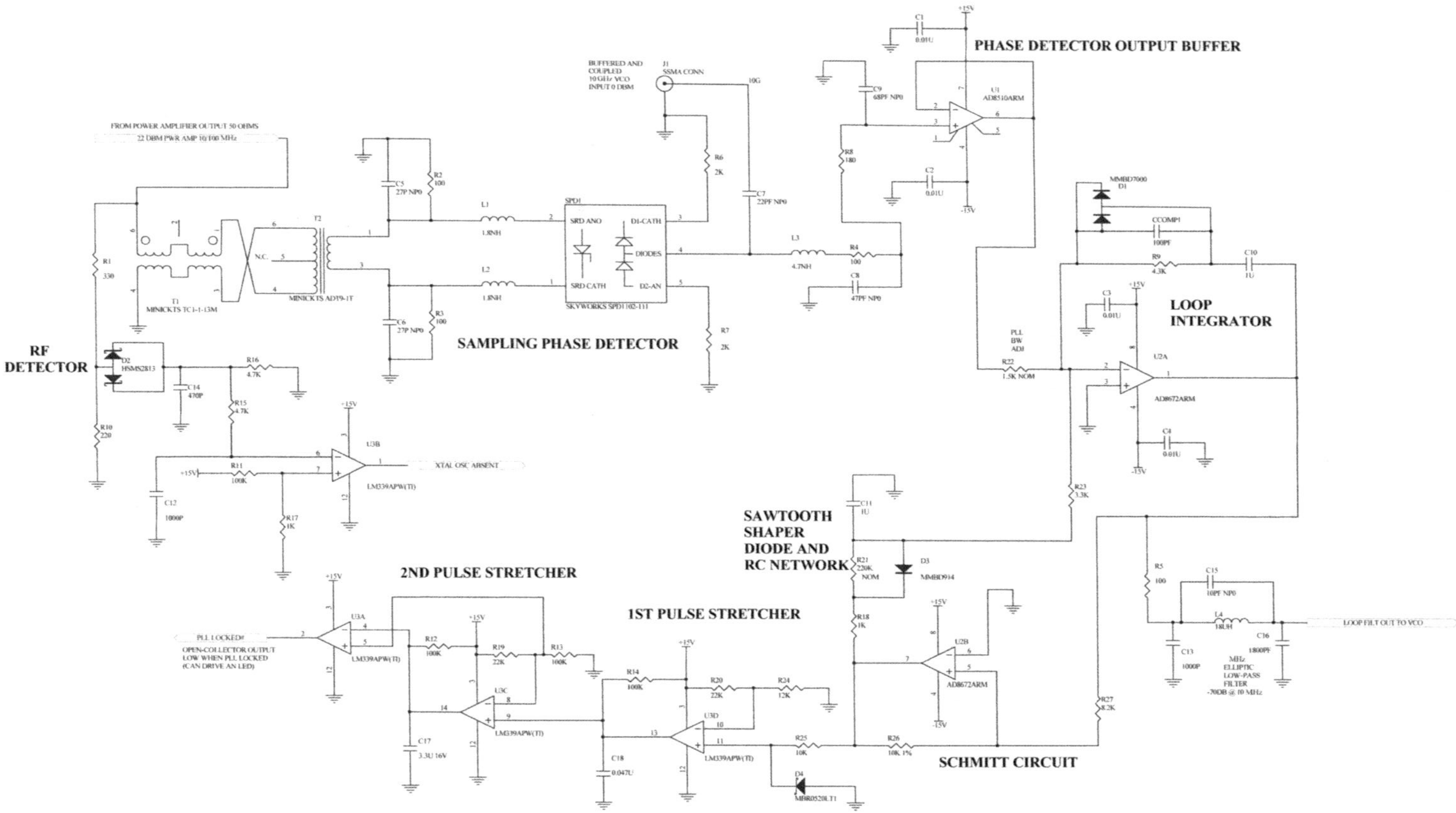

FIGURE 12.7 Self-sweeping PLL with sawtooth sweep characteristic using sampling phase detector to phase lock 10 GHz low-noise VCO to A 10 or 100 MHz low-noise quartz reference.

loop integrator. In this way, lock will never occur on the "flyback" slope of the sawtooth and will occur only on the slower ramp slope. The residual DC offset injected can be useful, as pointed out previously, in avoiding false locks due to sidebands. The circuit of Figure 12.7 can output a 10 GHz signal locked to a 100 MHz phase detector reference frequency without using a divide-by-*N* feedback counter, since the phase detector is an impulse sampler.

In a very low-phase noise application, we need to kill the noise coming from the Schmitt trigger. This is easily done by splitting the sweep current injection network and inserting a capacitor (C11) to ground. This also has the benefit of slowing the sweep and its discontinuity corners (beginning and ending the flyback state) so that the damping gain of the loop integrator (which is broadband) does not create step functions at the integrator output during the sweep peaks (diode D1 also shunts the damping resistor R9 to limit the magnitude of large step functions). This circuit can drive a DRO (dielectric resonator oscillator) or a VCSO (voltage-controlled surface-wave oscillator) having compatible tuning voltage profile (-11 to $+11$ V). The PLL shown was designed to have <200 Hz bandwidth.

Figure 12.8 shows a PLL using this self-sweeping technique as a simulation schematic. It omits the flyback portion of the sawtooth, so the sweep waveform is triangular, but it will illustrate how useful this approach is for both sweeping and avoiding false lock. We have superimposed a 1.1 MHz sine wave source ("spurious signal") at 0.4 V peak onto a primary sine wave of 1 MHz and 2 V peak. Thus, the spurious signal is one-fifth the amplitude of the desired signal. At the phase detector output, this ratio of 5:1 will be preserved for the error currents due to both competitors. If the maximum phase detector output peak error current due to the desired signal is 1 V divided by *R*, then we need a DC offset of 0.2 V divided by *R*. In Figure 12.8, we do not use currents but voltages, hence we set virtual $R = 1$. The Schmitt circuit input window is 0.5 times its +/− 10 V output rails making it change

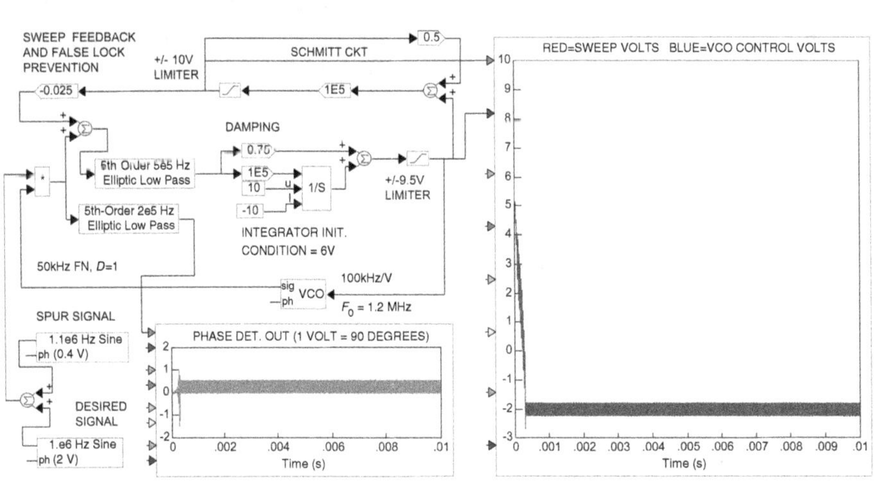

FIGURE 12.8 Automatic self-sweeping PLL having false lock prevention. (*See the color version of this figure in color plates section.*)

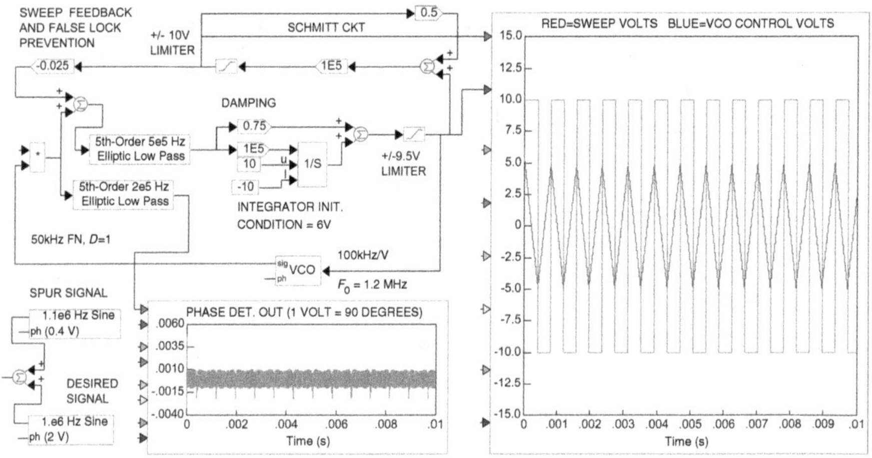

FIGURE 12.9 Automatic self-sweeping PLL having false lock prevention in no-signal sweeping mode. (*See the color version of this figure in color plates section.*)

state outside a +/ − 5 V input hysteresis. Thus, the VCO sweeps +/ − 5 V or +/ − 500 kHz. The Schmitt output rail of +/ − 10 V is weighted by 0.025 magnitude, resulting in +/ − 0.25 V of steady-state bias on the phase detector output, which the error value cannot overcome (negate) due to the spur. Proof is seen as the VCO begins its travel from 1.2 MHz downward, past the 1.1 MHz spur without pausing to lock, and finally settling at the desired frequency of 1.0 MHz. Notice the AC ripple on the VCO control voltage and on the phase detector output due to the spur 100 kHz away. All we have done is avoid locking on the spur, but its influence is still felt in the frequency domain as viewed on a spectrum analyzer. Since the spur had a value of −14 dBr (deciBels relative to a radian), it will appear as a pair of −20 dBc (decibels relative to the carrier) PLL output sidebands straddling the 1 MHz carrier on the spectrum analyzer assuming it is within the PLL bandwidth.

In Figure 12.9, we have disconnected all non-VCO signals driving the phase detector and now we can observe the unlocked PLL sweeping and attempting to find a frequency at which to lock. We can see the action of the Schmitt trigger circuit and its conversion from square wave to triangle wave by the loop integrator. In reality, the DC offset from the phase detector output is not zero. Often, it will exist due to second-order distortion of the VCO output waveform and asymmetry within the analog multiplier (diode imbalances). It is important to realize that a multiplier may rectify a nonsymmetrical input signal. The largest signal is usually the VCO output, thus this is the one to command scrutiny for symmetry of zero crossings. Also, the loop integrator op-amp may have DC offset, but it is usually minor compared to the sweep bias. The total DC offset determines the final phase angle to which the PLL will settle.

Figure 12.10 shows the acquisition and steady-state phase detector output DC (corresponding to a phase angle) when the PLL locks in the absence of a spurious

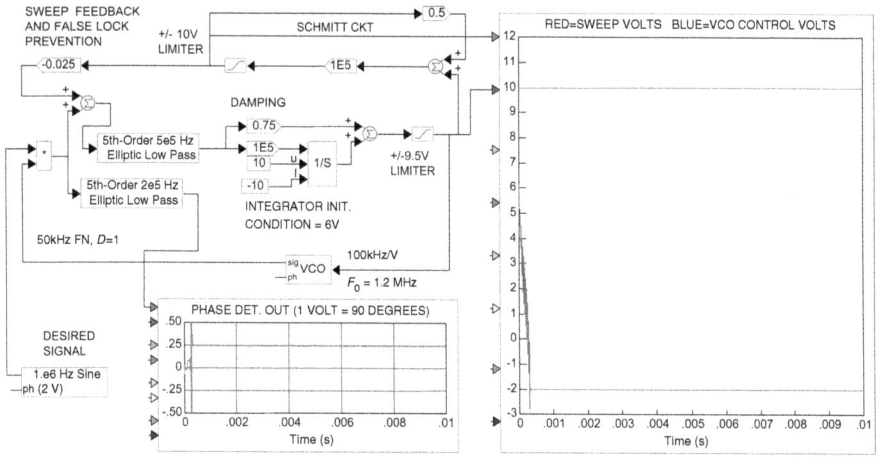

FIGURE 12.10 Automatic self-sweeping PLL having false lock prevention: no-spur behavior. (*See the color version of this figure in color plates section.*)

signal. Because the sweep is triangular, either polarity of DC offset from the phase detector is possible. Thus, steady-state settled phase is bivalued. To make it single valued, sawtooth sweep is required, as in Figure 12.7 and its narrative.

13

NONSWEEP ACQUISITION METHODS

13.1 DELAY LINE FREQUENCY DISCRIMINATOR

The delay line frequency discriminator offers many advantages. It has good linearity over a very wide frequency span. Its gain (volts per Hertz) can be designed to be very high or very low, as bandwidth is traded for gain. It involves few components, the major one of which is the delay line itself. Its output high-frequency ripple is at the carrier frequency or twice the carrier frequency and above. It can be used to demodulate the FM noise of the VCO and feed it back into the VCO out of phase, thus reducing the effective VCO's self-phase noise and also the total noise of the PLL. The delay line can be surface acoustic wave (SAW) type or coaxial cable.

The main drawback to its usefulness is its uncertain output voltage baseline due to lack of precision of the delay time, and temperature sensitivity of said delay, hence output baseline sensitivity to temperature. See Figure 13.1. An FM signal from an ideal VCO is applied to the delay line discriminator input, which consists of a normal path driving one input of a phase detector and a delayed path driving the other phase detector input. The delay line converts frequency to phase that the phase detector then converts to volts or amperes. Notice the ideally perfect conversion of delta frequency to delta volts. However, the delay line total delay usually has a tolerance of as much as 5%, making the final voltage baseline nonzero. The phase detector is assumed to be an analog multiplier with both inputs hard limited. Any non-50% duty factor of either the main or the delayed path will cause a gain uncertainty and possibly an output DC offset.

Frequency Acquisition Techniques for Phase Locked Loops, Daniel B. Talbot.

Published 2012 by John Wiley & Sons, Inc.

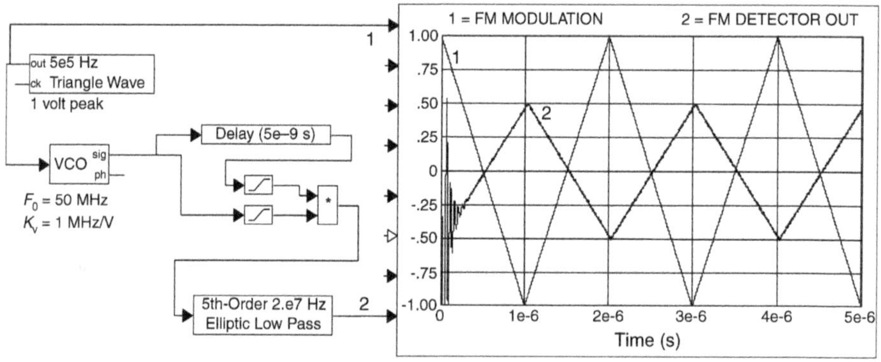

FIGURE 13.1 Delay line frequency discriminator centered at 50 MHz.

It should be used only in those cases where very large VCO frequency errors are expected compared to loop bandwidth. Or saying it mathematically,

$$\text{D}v >> v_n \tag{13.1}$$

where Dv is expected VCO frequency error and v_n is PLL closed loop natural frequency.

The toolkit VisSim file name is "Delay_Line_Freq_Discrim.vsm" and the reader is encouraged to "tweak the knobs" and alter the simulation to perform "what if" analysis. Typically, the delay line is set to exhibit one-quarter period of carrier delay, or some odd multiple thereof. The higher the integer multiple the higher the discriminator gain in volts per Hertz. Try it. In Figure 13.1, the delay has been chosen to be 5 ns, which is one-quarter of the period (quarter wavelength) of a 50 MHz signal frequency. While this results in very low gain, it does allow operation over extremely broad bandwidth. Try 15 ns and 25 ns and observe the output voltage scale increase. Finally, to see the output baseline intolerance to delay error, try 25.5 ns and notice the severe DC offset of the brown trace. This high offset for relatively minor delay errors makes the delay line discriminator unattractive except in software calibrated measurements or noise feedback.

A far more forgiving frequency discriminator is the quadricorrelator. Its disadvantage is its less than ideal linearity and frequency span capability and its output ripple frequency that is much harder to filter since it is twice the frequency delta from center. Its huge advantage is the absence of baseline DC output offset. Both the delay line discriminator and the quadricorrelator share the property of ignoring spurious sidebands unless they are larger than the main signal (per the capture effect explained in the previous discussions of encirclement noise) provided proper limiting is employed ahead of the frequency detection stage. The average DC output will be determined primarily by the main signal.

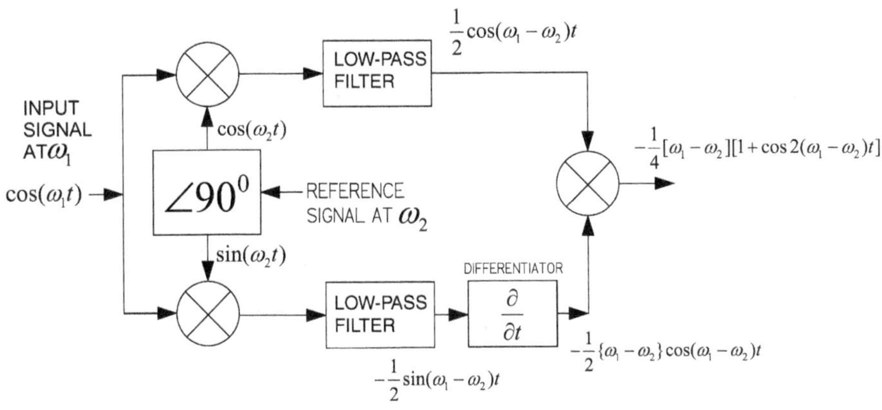

FIGURE 13.2 Unbalanced quadricorrelator as a frequency to voltage converter (discriminator).

13.2 THE FULLY UNBALANCED QUADRICORRELATOR

Refer to Figure 13.2. It shows the simplest form of a quadricorrelator. A frequency-to-voltage conversion is accomplished by quadrature signal processing to get I and Q baseband waveforms, then multiplying the I waveform with the derivative of the Q waveform (or by multiplying the Q waveform with the derivative of the I waveform). Notice that a ripple occurs superimposed on the output, the amplitude of said ripple being proportional to frequency error and the ripple frequency equal to twice the frequency error.

Mathematically, the output of the unbalanced quadricorrelator is written as follows:

$$E_{\text{out}} = -\frac{1}{4}[\omega_1 - \omega_2][1 + \cos 2(\omega_1 - \omega_2)t] \tag{13.2}$$

The unbalanced quadricorrelator is usually painful to use precisely due to its output ripple, although some applications do not care. Without much added complexity at all, (an additional multiplier and differentiator plus a subtractor) the unbalanced topology can be made balanced, and the objectionable ripple component canceled (of course, not perfectly, due to practical component tolerances).

The differentiator in the quadricorrelator is not strictly realizable (nor desirable for noise reasons), but is usually implemented with a single-pole RC high-pass filter having a corner frequency greater than the maximum expected frequency error. Because the high-pass filter flattens out at higher frequencies (whereas for true differentiation gain keeps increasing), the quadricorrelator output curve resembles an S-curve, similar to the old Foster–Seeley FM discriminator. Figure 13.3 is a VisSim simulation (in the toolkit as file name "quadricorr_unbal_sim.vsm") of the unbalanced quadricorrelator. Try "tweaking the knobs" (playing with the gains, filters, frequencies, etc.) to see what happens. Figure 13.4 shows the double-frequency ripple waveform superimposed on the frequency error waveform.

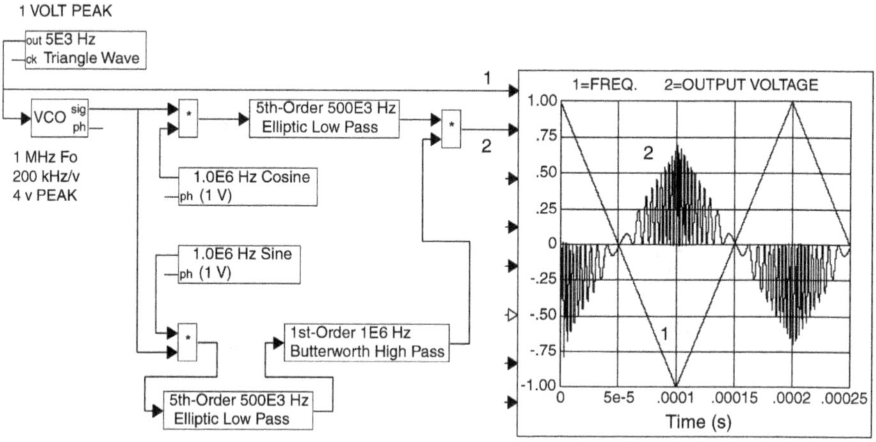

FIGURE 13.3 Unbalanced quadricorrelator output for a triangular frequency sweep (−200 kHz to +200 kHz).

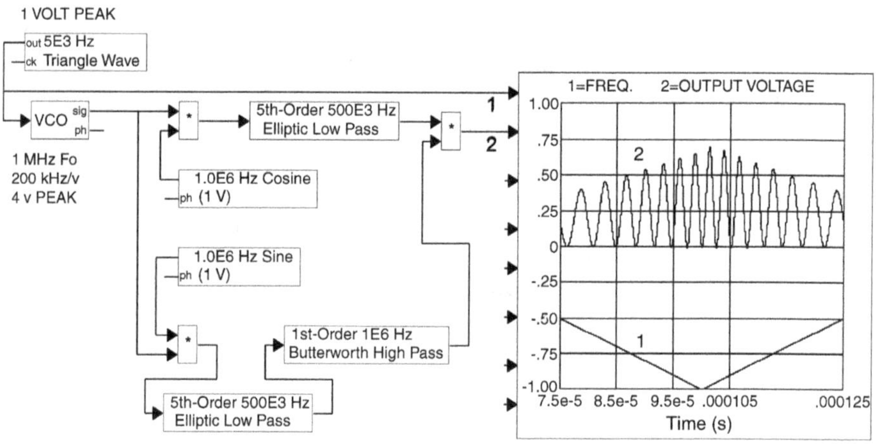

FIGURE 13.4 Unbalanced quadricorrelator ripple (at 200 kHz peak deviation it is 400 kHz, illustrating its 2F nature).

13.3 THE FULLY BALANCED QUADRICORRELATOR

While the unbalanced quadricorrelator's output contained ripple, the balanced version does not (or minimal ripple). The output for the unbalanced case can be expressed as

$$E_{\mathrm{OUT}} = Q\frac{\partial I}{\partial t} \tag{13.3}$$

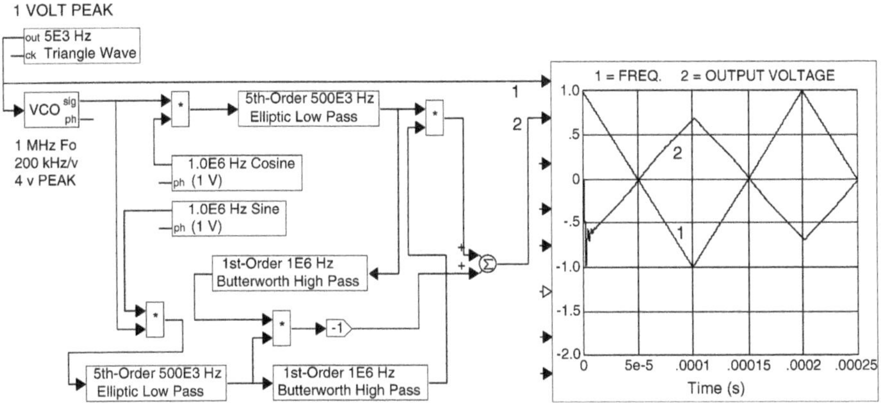

FIGURE 13.5 Balanced quadricorrelator output for a triangular frequency sweep of −200 kHz to +200 kHz (notice the absence of 2F ripple).

and the expression for the balanced case is given as

$$E_{\mathrm{OUT}} = Q\frac{\partial I}{\partial t} - I\frac{\partial Q}{\partial t} \tag{13.4}$$

and the VisSim simulation diagram is given in Figure 13.5. The VisSim file can be found in the toolkit as "quadricorr_bal_sim.vsm". Clearly, I and Q can be swapped in the above equations with only an output polarity inversion (at most) as a result.

13.4 THE MULTIPULSE BALANCED QUADRICORRELATOR

Occasionally, it is desired to measure the frequency of a repetitive pulsed signal. Such a signal has strong sidebands around the carrier, yet the quadricorrelator is not fooled. As we shall see, however, the quadricorrelator can be designed in a manner that makes this capability null and void, namely, by the incorrect application of sample-and-hold circuitry to increase the gain of the quadricorrelator for pulsed operation. Such an apparent gain benefit brings with it a nonsensical frequency to voltage conversion, unless implemented with matched filtering immediately prior to the sample and hold.

Figure 13.6 shows the output of a balanced quadricorrelator in response to a signal that is both pulsed with a low duty factor and swept in frequency (to show that there are no false conversions due to undersampling of the frequency offset).

On the other hand, Figure 13.7 shows what happens when the brief output pulses are fed through sample-and-hold(s). Due to the presence of filter ringing (and, in the case of placement before the subtractor, the 2F ripple), the staircase waveform that results is totally inaccurate.

However, when each quadricorrelator output pulse is averaged by matched filters before application to the sample–holds, the filter artifacts and ripples are largely ignored, as shown by the resulting orderly and accurate staircase waveform of Figure 13.8.

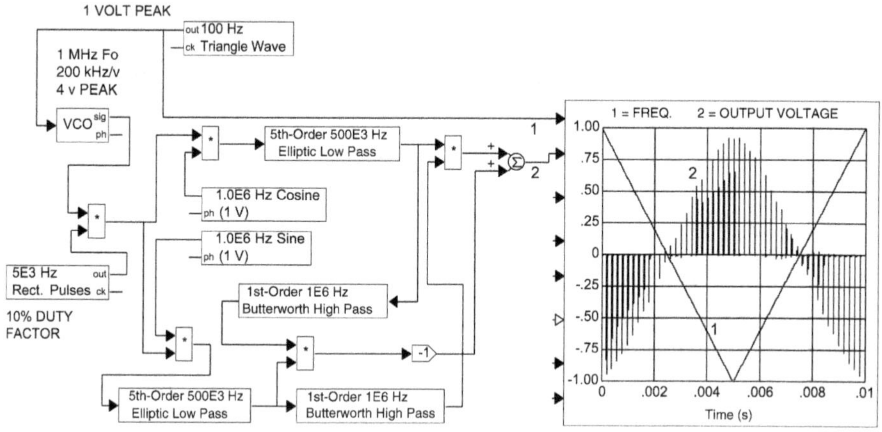

FIGURE 13.6 Balanced quadricorrelator output for a pulsed triangular frequency sweep of −200 kHz to +200 kHz (notice the absence of 2F ripple).

The proper staircase waveform from a pulsed frequency sweep is shown in Figure 13.9 using just one matched filter followed by just one sample–hold. Each step in the staircase is caused by a single frequency value as shown by the sweep waveform, which represents the frequency of the swept signal versus time. Notice that the sweep rate is exceeded by the pulse rate. Also notice that the frequency offset swings by up to +/− 50 kHz peak deviation but the pulse rate is lower than that (10 kHz). In most heterodyne or homodyne systems, this would result in aliasing, rollover, or foldover of the staircase waveform, but the quadricorrelator when properly implemented escapes the problem.

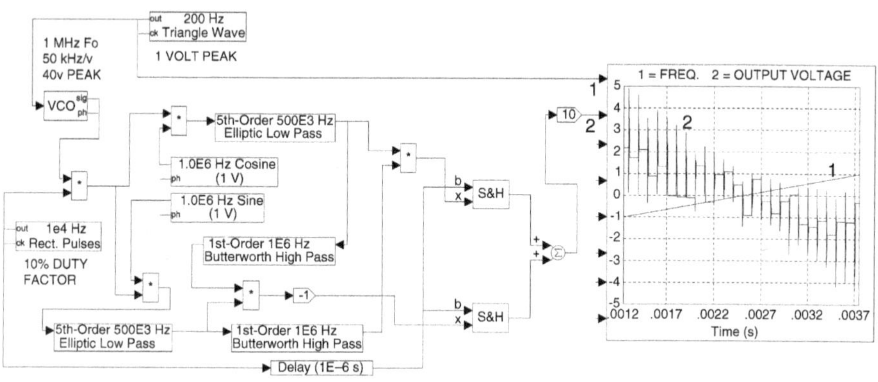

FIGURE 13.7 Multipulsed balanced quadricorrelator with sample–holds, no matched filters (notice uneven false stairsteps).

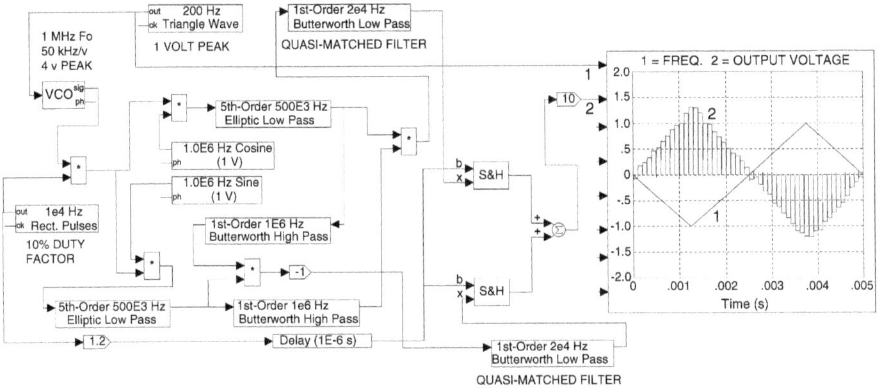

FIGURE 13.8 Balanced quadricorrelator output for a pulsed triangular frequency sweep (−50 kHz to +50 kHz) with quasi-matched filters and sample–holds.

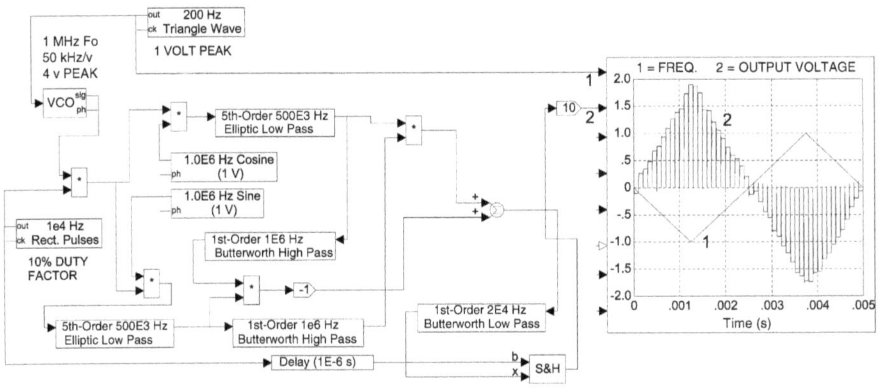

FIGURE 13.9 Multipulsed balanced quadricorrelator with one matched filter and sample–hold (notice proper stairsteps).

13.5 CONCLUSION REGARDING PULSED FREQUENCY DETECTION

The use of either a delay line or Foster–Seeley or quadricorrelator frequency discriminator carries a common benefit: immunity from false frequency identification unless there exists interference greater than the carrier magnitude, or the quadricorrelator is improperly designed, or the interference is outside the linear frequency span of the quadricorrelator. The high-pass filter employed in the quadricorrelator is only an approximation to a true differentiator. Above the high-pass filter's cutoff frequency and somewhat below that, its phase angle departs from 90°. To the degree it does, the quadricorrelator output versus frequency will bend from a straight line. In doing so, the beat waveform (which we have seen to be nonsinusoidal) will be altered (distorted) and thus DC offset will occur. Of course,

DC offset can also occur from more pedestrian causes, such as multiplier nonlinearity and offsets. The designer must trade off linearity over a given frequency span for conversion gain-to-offset ratio. Like the delay line discriminator, trying to achieve good linearity over a wide frequency span results in low output amplitude that does not compete well with output DC offsets. Unlike the delay line discriminator, the output offset is much lower for similar component tolerances.

The DC offsets of both the delay line discriminator and the quadricorrelator can be tuned out using software calibration, provided it is updated adequately often, at the expense of more complexity (circuitry to interface to the software). For frequency acquisition purposes, the DC offset of either approach is more interesting (and more challenging) than the conversion gain (volts per Hertz). A variation in conversion gain is well tolerated by the frequency acquisition process, since what is primarily desired is frequency error polarity and center information. The conversion gain of the frequency discriminator affects acquisition time in about direct proportion for a first-order frequency loop. In other words, if the discriminator gain drifts from 1 V per MHz to 1.2 V per MHz, we will see perhaps 20% faster frequency convergence before the PLL locks. However, if the DC baseline of the discriminator changes, it must be such that the converged frequency error can be "eaten" by the PLL. After all, the PLL capture range without frequency acquisition assistance is finite (unless nearly infinite settling time is allowed), and if the frequency loop settles to a value outside this capture range, the PLL may be pulled off center so that it can never capture.

Recall that one of the main reasons for using outside assistance for the PLL frequency acquisition is that the unassisted PLL may take a long time to generate enough of its own DC error signal to lock quickly if the beat frequency is high enough to be strongly attenuated by the loop filter. While the type-II PLL can in theory generate a DC error from the distortion of the beat waveform sufficient to drive it toward lock, this error may be so feeble (due to roll-off in the loop filter) as to not compete with DC offsets and noise, or may be so feeble that it would take years to lock in the absence of those offsets and noise. Thus, we need frequency acquisition assistance. Most PLL IC designs achieve this by using the edge-triggered PFD, but that buys the noise problems and inability to handle missed edges previously discussed. But we will concede that often this degraded performance is acceptable. When it isn't, all of our previous discussions and emphasis are pertinent.

13.6 QUADRICORRELATOR LINEARITY

See Figure 13.10. If the "differentiators" in the balanced quadricorrelator are implemented using single-pole high-pass filters having inadequately high corner frequencies, a nonlinearity in output will occur. Note in curve 1 the input frequency normalized to -1 to $+1$ volt (representing -200 kHz to $+200$ kHz) and curve 2 the corresponding output voltage from the distorted quadricorrelator. Instead of obeying a straight line from zero frequency offset to the frequency extremes, the actual output curve bends into a nearly "S"-shaped curve. This is not an unusual curve for those acquainted with the Foster–Seeley and ratio detector type of frequency discriminators.

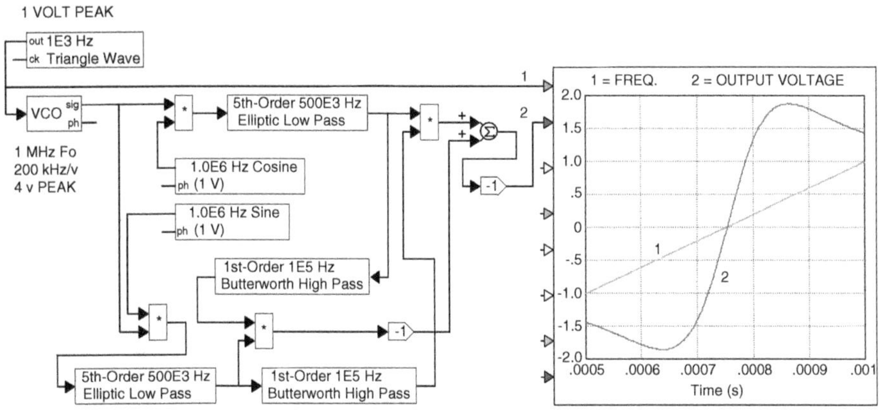

FIGURE 13.10 Balanced quadricorrelator without and with S-curve shape.

Note that because of this S shape, the output has two slopes. Some may be alarmed by this, but it is not a problem for loop stability. This is because the slope reversal occurs superimposed on the correct absolute polarity. Specifically, the curve for negative frequency offsets remains the correct polarity and for positive frequency offsets also remains the correct polarity. Of course, we did not plot the curve beyond +/−5 times the high-pass corner frequency. Well, beyond those limits it may pose closed loop problems. But in general, one may employ the distorted quadricorrelator without anxiety. Notice that there is a benefit of using it: the slope of the curve as it passes though zero is steeper, indicating more sensitivity in volts per Hertz, which is beneficial in overcoming DC offsets.

In Figure 13.11, the architecture has been altered to show the effect of unequal differentiator corner frequencies. It has been exaggerated to 2:1 inequality for

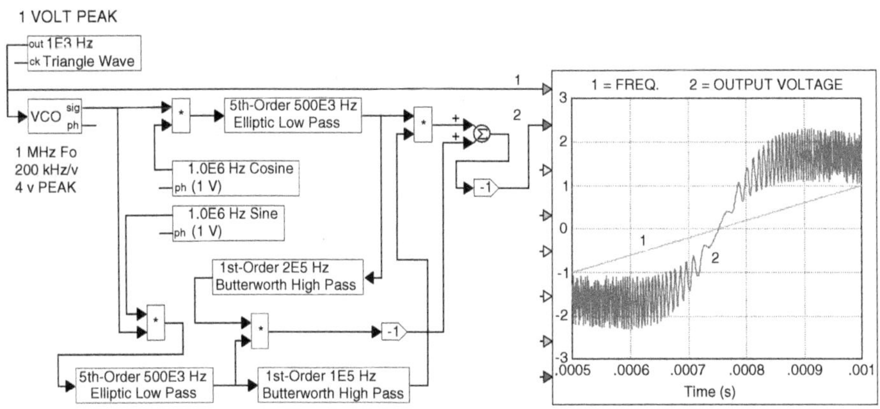

FIGURE 13.11 Unbalanced differentiators in the quadricorrelator result in ripple.

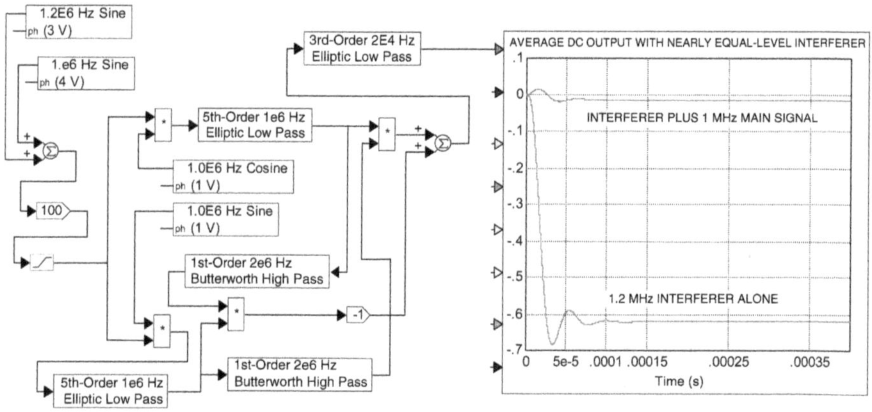

FIGURE 13.12 Balanced quadricorrelator output without S-curve shape and 200 kHz interfering sideband and heavy limiting.

purposes of example. The result is markedly increased ripple. In practice, much smaller inequality will occur due to component tolerances.

The effect of an S-curve on interference susceptibility is also easily simulated and its impact depends on interferer frequency relative to desired frequency as the loop acquires frequency lock. (*Note*: the interferer may be varying in frequency as the loop closes.) See Figure 13.12 and the VisSim file in the toolkit, file-named "quadricorr_bal_with_S-CURVE + SIDEBAND_sim.vsm." In this example, hard limiting is employed, with the degree of clipping (which we shall name "depth of limiting") being 100:1. By that we mean a gain of 100 exists prior to the symmetrical clipper having output rails equal in amplitude to the amplitude of the main signal. Note that the interfering signal is 200 kHz away, and is 0.75 times the main signal's amplitude (recall that if $A_\mathrm{i} > A_\mathrm{m}$, the interfering signal in essence becomes the main signal and the main signal becomes the interferer, as explained in our previous discussions of the FM capture effect, also called vector encirclement, where A_i is the amplitude of the interferer and A_m is the main signal amplitude). Notice the fairly good rejection of the interferer when present with the main signal (only about 15 mV departure from zero). Also notice that when the main signal is removed, the interferer becomes in essence the main signal, and generates about 0.63 V per 200 kHz offset (the quadricorrelator center frequency is 1 MHz corresponding to 0 V). This quadricorrelator has high-pass corners equal to 10 times the interferer offset frequency.

Now, consider the case of Figure 13.13, where the quadricorrelator high-pass filters have corners to the interferer offset frequency (200 kHz), with the same depth of limiting. The gain of the quadricorrelator is now much greater but with more of an "S" shape with the knees of the "S" occurring near +/−200 kHz. Also notice a high degree of immunity to the influence of the interferer when the main signal is present.

Of course, we would hope that the amplitude of the interferer is much less than 0.75 as in this exaggerated example. But what if the depth of limiting is reduced to 1:1? If limiting depth is low, some AM component will coexist with the PM vector

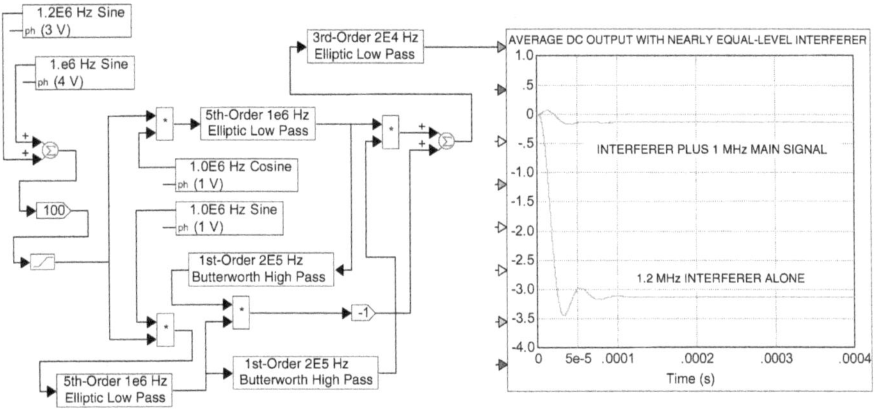

FIGURE 13.13 Balanced quadricorrelator output with S-curve shape and 200 kHz interfering sideband and heavy limiting.

and this AM will not be fully ignored by the quadricorrelator. Thus, it is vital that if one expects to encounter spurious sidebands around the main signal frequency, one need employ a hard limiter having a high depth of limiting. Just see what happens to our perfectly good quadricorrelator example when depth of limiting is reduced to unity. This undesirable condition is illustrated in Figure 13.14. The interferer is now able to generate an output voltage absolute value much greater than zero. This bodes badly for frequency convergence in the loop.

Figure 13.15 shows an ideal FM detector (delay line frequency discriminator) output under similar conditions at various depths of limiting, but with an additional problem inserted into the scenario, namely, a DC offset prior to the limiter. We have chosen to make this DC offset fluctuate by modeling it with a low-frequency sine

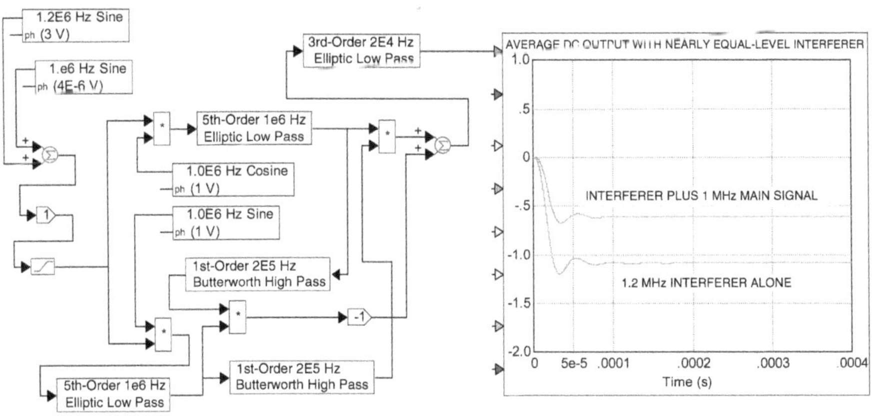

FIGURE 13.14 Balanced quadricorrelator output with S-curve shape and 200 kHz interfering sideband and only very mild limiting.

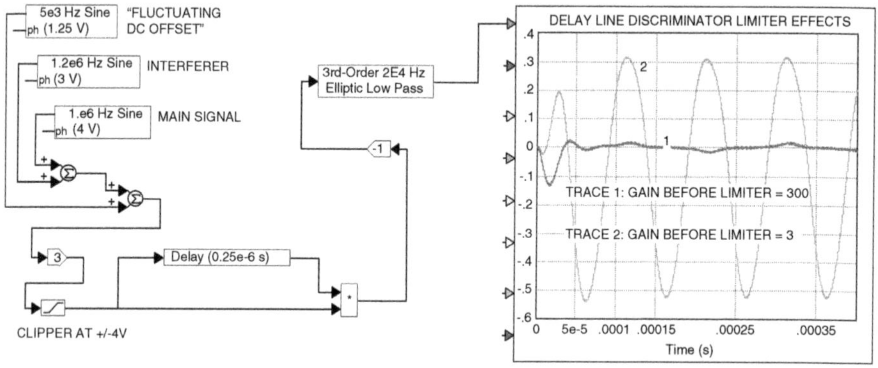

FIGURE 13.15 Response of delay line discriminator to interferer as a function of depth and symmetry of limiting.

wave. Of course, in any limiter there will in general be a DC offset. This skews the limiter centroid, causing asymmetrical limiting and consequentially AM-to-PM conversion. Before we illustrate AM-to-PM conversion, notice that when depth of limiting is high, more immunity to DC offset (more limiter asymmetry) is tolerated.

Figure 13.16 shows how depth of limiting affects the ideal delay line discriminator without any asymmetry. Clearly, the reduction in depth of limiting from 300 down to only 3 has very bad consequences. But notice the fact that in the presence of a spurious interferer at high depth of limiting, the ideal delay line discriminator output is closer to zero volts than our not-so-perfect quadricorrelators. This proves that FM capture is possible and has indeed occurred, and that the effectiveness of the capture effect is indeed a function of depth of limiting. The reader may wish to run the VisSim simulations again using his or her own numbers for depth of limiting, but it

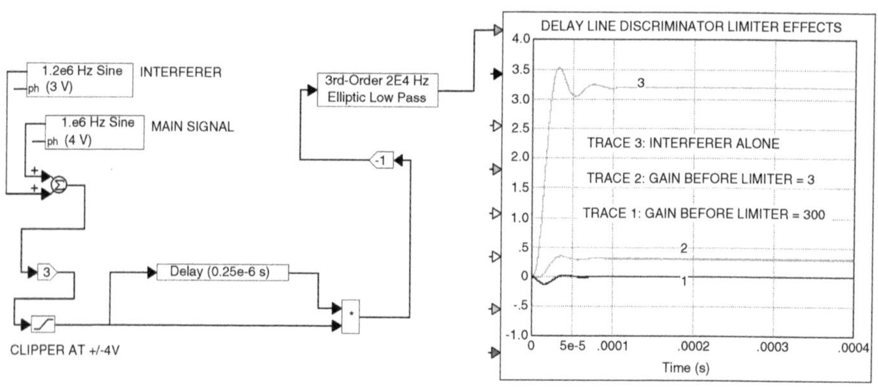

FIGURE 13.16 Response of delay line discriminator to interferer as a function of depth of limiting only.

appears that diminishing benefit occurs for 100:1 or greater limiting depth. If the reader has skipped the chapter discussing encirclement vectors and FM capture, we recommend reading it now for a better appreciation of these phenomena (depth and symmetry of limiting). In that discussion, it was given as a condition that the ensemble vector phase was the variable of interest in the subsequent equations. If limiting is omitted or too soft, it is clear that the resultant ensemble vector has both amplitude modulation and phase modulation. The response of a given topology of frequency detection scheme to this AM will depend on its AM rejection behavior. We will explore the AM rejection of a particular quadricorrelator and compare it with that obtained using the ideal delay line discriminator a little later.

13.7 LIMITER ASYMMETRY DUE TO DC OFFSET

In Figure 13.17, a typical limiter is shown having a model for DC offset between its two inputs. This offset can be made very small by matching the two transistors. A perfectly matched pair of transistors will generate a symmetrical output waveform when driven by a sinusoid, as shown in Figure 13.18 (assuming a properly symmetrical differential amplifier at high frequencies where it is used) using a low-frequency 1 kHz sine wave for illustration. Such a symmetrical waveform will have no even harmonics (see spectrum of Figure 13.19). But if a DC mismatch exists between the two halves of this differential configuration, even order harmonics will begin to appear (see Fig. 13.20). Observing the output spectrum may, therefore, prove to be a more sensitive metric of DC offset than direct DC measurements. For a BJT (bipolar junction transistor), the relationship between output and input for a single-ended grounded emitter BJT is 3 mV of input for each dB of output. Thus, if we want 40 dB (100:1) depth of limiting, we need an input swing of 40 times 3 mV or

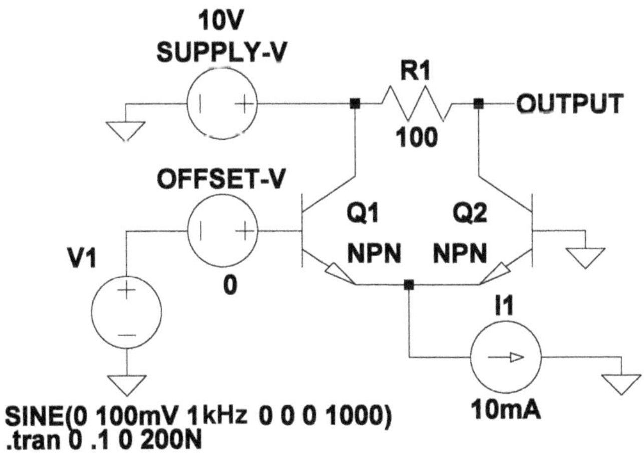

FIGURE 13.17 BJT hard limiter to show effects of DC offset.

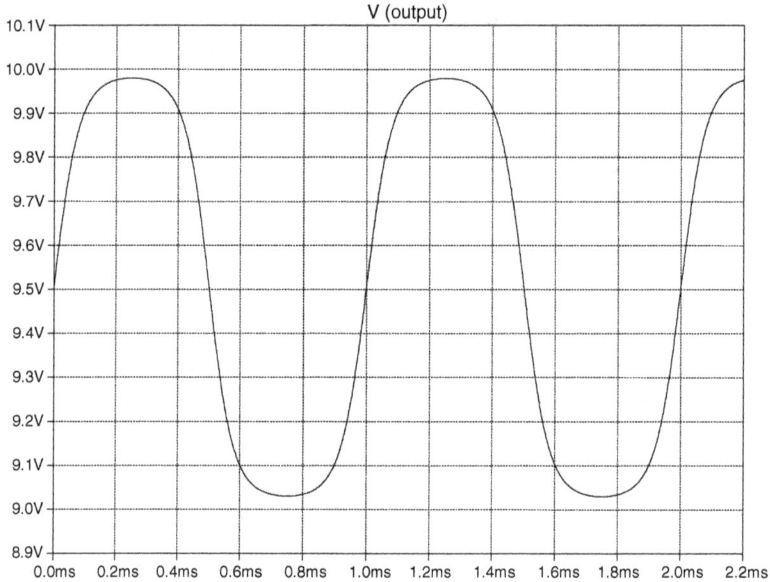

FIGURE 13.18 Output of symmetrical limiter, no DC offsets.

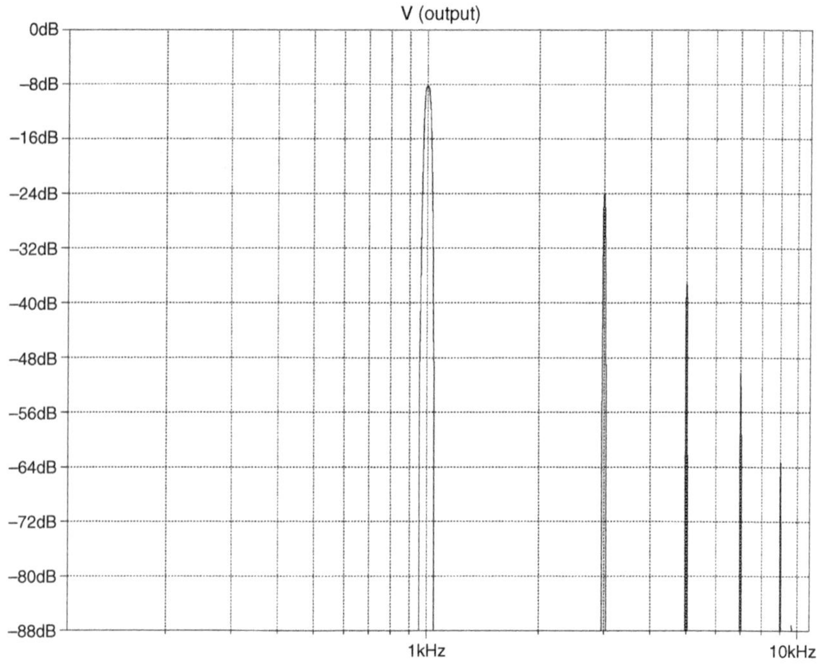

FIGURE 13.19 Output spectrum of symmetrical limiter (no DC offset) shows only odd harmonics (absence of second-order distortion).

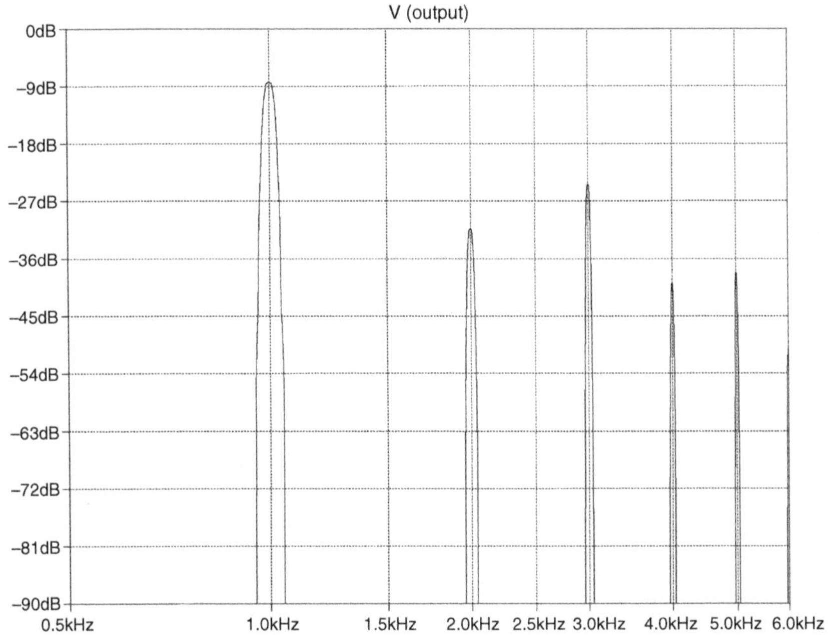

FIGURE 13.20 Limiter output spectrum with 10 mV input DC offset showing the presence of second-order distortion.

0.12 V *peak*. Of course, a single-ended transistor makes a horrible limiter since it is by definition not symmetrical in either topology or behavior. But it helps to understand that for our symmetrical limiter, a base-to-emitter mismatch of only 3 mV between the two BJTs is fairly trivial, but if discrete BJTs were used with no thermal tracking between them, 10% asymmetry could easily occur over a significant temperature variation. In any event, DC offset can be nulled with a trimpot while observing even-order harmonics by adjusting the trimpot for minimum second and fourth harmonics. (Alternatively, AC coupling between the two emitters can be used to eliminate the DC mismatch from AC consideration. However, IC designers may not be able to incorporate such a low-impedance capacitor on-chip.)

It might be instructive to run the LTspice simulation the author has included in the folder called "Spice files from author" containing the file named "BJT_HARD-LIMITER_WITH_ASYMMETRY.asc." Change the input source frequency from 1 kHz to 100 MHz, add Miller capacitance from collector to base of Q2 of 1 pF, then add a resistor in series with the base of either half of the transistor pair and observe its impact on symmetry (even harmonics will appear) even when no DC offset is present. This phenomenon is termed "diagonal clipping" or AC asymmetry. The same will occur if power supply voltage is reduced to 1 V or so. The punch line is that for good symmetry at low power supply voltages, we need very fast and well-matched transistors if the simple topology of Figure 13.17 is employed. A more symmetrical topology is readily achievable. Operational amplifier designers will have an image come to mind, but remember that at high frequencies, complexity breeds impaired performance.

The converse relationship of even-order harmonics to DC offset is also possible: DC offsets are always generated as an accompaniment to even-order distortion in any circuit. This is true when the circuit transfer function is expanded in a Taylor series to show that the byproduct of second-order (or any even order) signal distortion is a DC offset.

13.8 TAYLOR SERIES DEMONSTRATES SECOND-ORDER-CAUSED DC OFFSET

A linear amplifier or mixer or A/D or D/A converter (etc.) is desired to have a linear relationship between its output and input. Linear means that if we double the amplitude of the input, the output will exactly double. In reality, this almost never occurs. The output may increase slightly more than double (expansion) or slightly less than double (compression). Furthermore, the departure from linearity generally gets worse as the signal amplitude increases or decreases. The Taylor series helps us understand this departure from a straight line.

First, the equation for a linear amplifier or other circuit can be written as

$$E_{\text{out}} = K_1 E_{\text{in}} \tag{13.5}$$

where E_{out} is the circuit output voltage, E_{in} is the voltage applied to the circuit input, and K_1 is the linear gain of the circuit.

If we were to graph the output versus input voltage, it would be a straight line on a linear X and Y plot.

But in practice, there is no such animal as a linear circuit except for passive components and even then there is nonlinearity. Consider, for example, the nonideal resistor. Doubling the voltage across it should double the current through it, but in reality it will heat up a little bit, which will alter its resistance, so the current will not exactly double.

So we need to represent the real world by a more complicated equation, one that includes several forms of departure from a straight line. One form of departure is second-order, that is, a squaring of the input signal diluted by some constant K_2. We then superimpose this departure upon our original linear behaving equation to get a second-order contribution. In a similar way, we weight a third-order term by a third-order coefficient and add it (superposition) to the overall equation, to get a Taylor series that now looks like this:

$$E_{\text{out}} = K_1 E_{\text{in}} + K_2 E_{\text{in}}^2 + K_3 E_{\text{in}}^3 + \cdots \tag{13.6}$$

where the . . . indicates continuity of format to fourth, fifth, sixth, seventh, and so on order terms.

Most circuitry conforms closely to the above behavior for weak nonlinearities. A more exact behavior is predictable using the Volterra series [26,27] incorporating memory effects.

It is not difficult to see that the term $K_2E_{\text{in}}^2$ generates a DC term for any polarity of E_{in}.

Likewise, it is not difficult to see that the third-order term does not add a DC component to E_{out} but does add a small copy of the original signal due to the fact that the third-order term is basically a second-order term (generating DC) multiplied by a first-order term. This small added fundamental copy either increases the output fundamental signal amplitude (if K_3 is positive) or decreases it (if K_3 is negative) corresponding to expansion or compression, respectively.

Thus, DC is always produced (semirectification) when second-order nonlinearity is present, although it may be subsequently AC coupled so that the observer does not see it. This suggests that a counteracting DC term may be superimposed *on the input signal* to modify the other second-order *output* products such as the second harmonic that is produced by squaring a cosine wave representing E_{in}. So far, we have been assuming only one signal for E_{in}, namely, a sine wave. What if two signals at different frequencies make up the input voltage E_{in}? This is called a two-tone condition. We would represent E_{in} by

$$E_{\text{in}} = \alpha\cos(\omega_1 t) + \beta\cos(\omega_2 t) \tag{13.7}$$

where the first term to the right of the equality is a cosine wave of frequency ω_1 and amplitude α and the latter term is a cosine wave of frequency ω_2 and amplitude β.

If this two-tone waveform is plugged into Equation 13.6, the output will consist of several harmonics and cross-products. We will consider those cross-products later, but we will simply say now that they are usually more problematic than the individual harmonics in most communication systems and circuits.

13.9 THIRD-ORDER INTERMODULATION DISTORTION AND TAYLOR SERIES

When the signals of Equation 13.7 are applied to a circuit having distortion modeled by the Taylor series approximation of Equation 13.6, several new signal frequencies are produced.

Let us use standard frequency in Hertz rather than radians/second $[f = \omega/(2\pi)]$. Then, our two original signal frequencies are f_1 and f_2. The second-order exponent term in Equation 13.6 gives rise to DC as previously mentioned, plus new signals at $(f_2 - f_1)$, $(f_2 + f_1)$, $(2f_2)$, and $(2f_1)$. Clearly, all of these signals are far away from our original signals. Unfortunately, the third-order exponent term causes new signals to also appear at $(3f_1)$, $(3f_2)$, both of which are far away, and at $(2f_2 - f_1)$ and $(2f_1 - f_2)$. These last two frequency signals are called "third-order intermods" and are close to our desired original signal frequencies, thus difficult or impossible to remove with nonheroic filtering in general. They are "in our face," or "in-channel" or "in-band" in most systems. Their amplitudes are all predictable from the Taylor series. When the

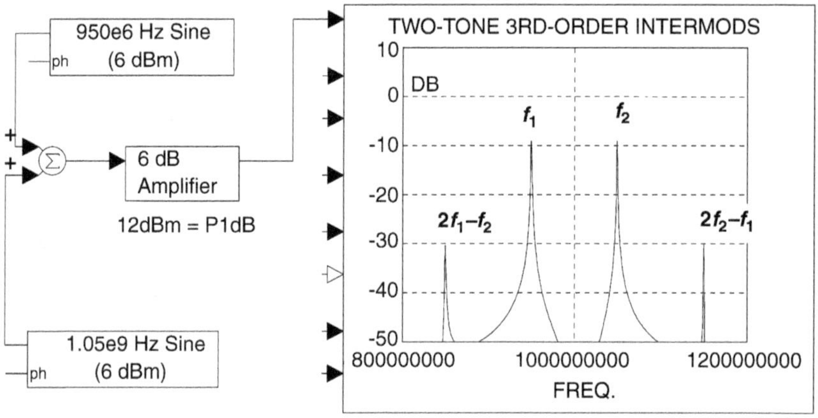

FIGURE 13.21 Third-order intermodulation products.

two tones are equal in amplitude, they produce the third-order intermods as shown in Figure 13.21.

Notice that third-order products all vary in amplitude as the cube of the input amplitude varies. Thus, if we drop the amplitude of the input signal(s) by 1 dB, these third-order terms drop by 3 dB. The relative difference is, therefore, 2 dB, which is very important. It tells us that for every 10 dB that we dilute the input signal, there is a relative benefit of 20 dB in the reduction of third-order products. Similarly, the second-order terms drop twice as much in dB as the drop in input level, which is difference between the main signals and the second-order distortion. Thus, if we drop input two-tone levels by 10 dB, our distortion (as a fraction of main signal levels) improves only 10 dB (still an improvement but not as dramatic as that for third-order distortion).

There is a technique for quantifying the third-order distortion one will encounter when operating a device from a given manufacturer at the user's desired amplitude. It is known as the third-order intercept. It is fictitious point on a graph below which one obtains the two-for-one decibel improvement relative to 0 dB. For example, if an amplifier (or other device) is specified to have a third-order intercept point of 30 dBm, then backing down to 0 dBm gives us −60 dB third-order distortion relative to our dropped input level. Similarly, for second-order distortion there is a second-order intercept specification from the manufacturer. If it is 25 dBm, then backing down to 0 dBm improves our second-order distortion terms to −25 dBc or so. For a more thorough discussion of third- and second-order intercept and 1 dB compression (also a useful point that occurs about 13 dB below the third-order intercept point), see Refs [28–30], and many others too numerous to list.

The equation for the amplitude E_3 of third-order distortion products for m number of inputs is

$$E_3 = (3/4) \sum_{N=1;P=1;N\neq P}^{m} K_3 E_N{}^2 E_P \cos(2\omega_N \pm \omega_P)t \tag{13.8}$$

where m is the number of input signal frequencies, and the result will be $2m(m-1)$ frequencies at which third-order intermods will occur.

This discussion is not only very useful for predicting intermods but is also useful as an instrumentation diagnostic in the design lab. If one notices a spurious signal, one can decide whether it is really present or is instead an artifact of one's lab instruments. For example, if a spur exists at −20 dBc and adding a 6 dB attenuator in front of the spectrum analyzer causes it to drop by 18 dB (by 12 dB with respect to the main signal that has, of course, dropped by 6 dB, making new observation −32 dBc), then the spectrum analyzer was being overdriven and the spur may not be as bad as first observed (assuming only third-order compression in all affected devices).

14

AM REJECTION IN FREQUENCY DETECTION SCHEMES

14.1 AM REJECTION WITH LIMITER AND INTERFERER

In both the delay line discriminator and the quadricorrelator, it is easy to see that, without a preceding limiter, a square-law relationship exists between signal amplitude at the input and conversion gain (volts per Hertz). In the delay line discriminator, the delayed and nondelayed paths are multiplied, so the squaring relationship is obvious. Similarly, in the quadricorrelator, the amplitude of Q or I is proportional to input signal amplitude, and thus Q multiplied by the derivative of I or vice versa results in squaring. This squaring is maintained both in the balanced and in the unbalanced quadricorrelator topology. Thus, one kind of AM "nonrejection" is already identified in the absence of interfering sidebands or spurs. Without a limiter preceding the frequency-to-voltage converter, there is no rejection of AM; in fact, AM is squared.

But less obvious is the result of AM in the presence of spurs prior to a limiter preceding the frequency detection. In this case, AM is equivalent to time-varying depth of limiting. For our version of AM, we will amplitude modulate both the interferer and the main signal simultaneously, preserving their ratio but effectively varying the dept of limiting. We choose to vary the multiplication of gain prelimiting by 0–1. Since we have used a prelimiter gain of 30, the effect will be to vary the limiting depth from 0 to 30. We have also decided to amplitude modulate using a low-frequency sine wave. The results underscore the need for high corner frequency in the low-pass filters of the quadricorrelator compared to the maximum expected

Frequency Acquisition Techniques for Phase Locked Loops, Daniel B. Talbot.

Published 2012 by John Wiley & Sons, Inc.

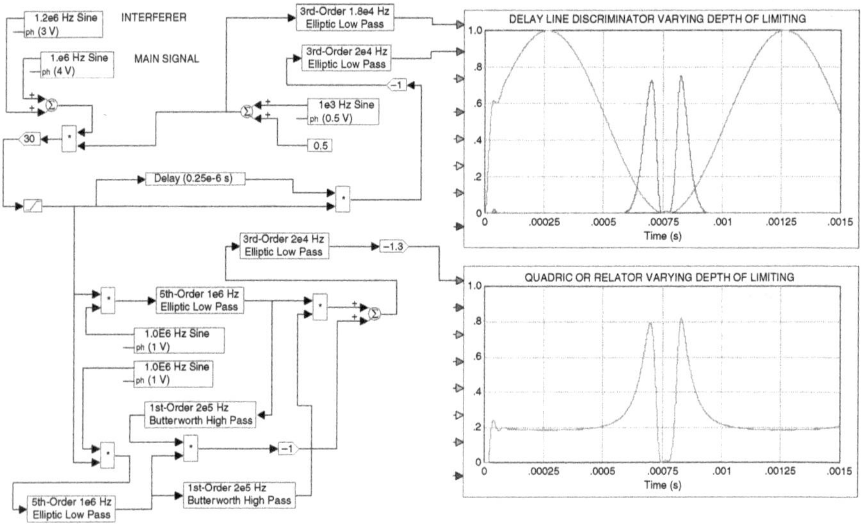

FIGURE 14.1 AM rejection with S-curve quadricorrelator. (*See the color version of this figure in color plates section.*)

interferer offset frequency. We begin our simulations by showing the modulation (in the top traces of the following figures) normalized to unity peak sine wave amplitude (corresponding to depth of limiting of 30). This AM sine wave is obtained by adding a DC offset of 0.5 to a sine wave of 0.5 V peak, then multiplying the signal fed to the limiter by this sum times 30. Refer to Figure 14.1.

14.2 AM REJECTION OF THE BALANCED LIMITER/QUADRICORRELATOR VERSUS THE LIMITER/DISCRIMINATOR IN THE PRESENCE OF A SINGLE SPUR

In Figure 14.1, we have amplitude modulated the depth of limiting by 0–100% of 30. It was found that little benefit accrues for depths greater than 30. An interferer spaced 200 kHz away from the main carrier of 1 MHz was added at three-fourths of the amplitude of the main carrier. This is an interferer of −2.5 dBc that is quite a large spur. From Equations 8.6 and 8.7, we saw that the interferer generates a distorted beat waveform at the frequency discriminator output, and that this waveform grows in amplitude and distortion asymmetry as the amplitude of the spur increases, and grows in direct proportion to spur offset frequency from the main carrier. Thus, the designer should strive for suitable performance under conditions of maximum expected spur offset frequency and amplitude. The cusps of the distorted beat waveform (see Figure 8.6) should not be clipped; otherwise, DC baseline shift will result, yielding a shift in apparent measured center frequency even though no such shift existed. In that case, the frequency acquisition will converge to the wrong frequency, quite probably preventing PLL

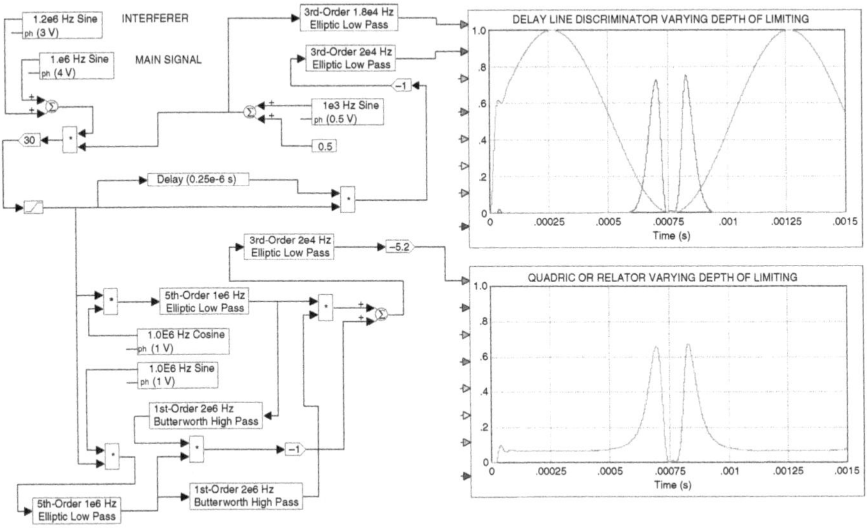

FIGURE 14.2 AM rejection by extending the S-curve width. (*See the color version of this figure in color plates section.*)

acquisition (which was the objective of the frequency steering provided by the frequency detector, be it quadricorrelator or other). Notice the following observations from Figure 14.1:

(a) When the modulation goes to zero depth (at 750 μs), the outputs of both the limiter/discriminator and the limiter/quadricorrelator go to zero; this is intuitively obvious.

(b) When the modulation goes to 100% (depth of limiting = 30:1), the limiter/discriminator goes to very nearly zero also (at about 250 and 1250 μs), but the quadricorrelator (lower trace) goes to 0.2 rather than zero for the condition that the differentiator corner frequencies are equal to the interferer frequency offset (spacing from main carrier).

Thus, a high-gain quadricorrelator having pronounced "S-curve" shape close to our interferer offset is not very good at rejecting the interferer's influence even at heavy limiting prior to the quadricorrelator.

Now observe Figure 14.2. We have maintained all quantities the same except that we have moved the differentiator corner frequencies to be 10 times the interferer offset and adjusted the conversion gain to be the same as in Figure 14.1. Notice the definite improvement over the previous case in the lower trace at 100% limiting.

Next, observe the fact that if we alter the differentiator corner frequency to be 100 times the interferer offset frequency and renormalize the conversion gain, no noticeable improvement is obtained (see Fig. 14.3). Thus, the factor 10 is probably sufficient.

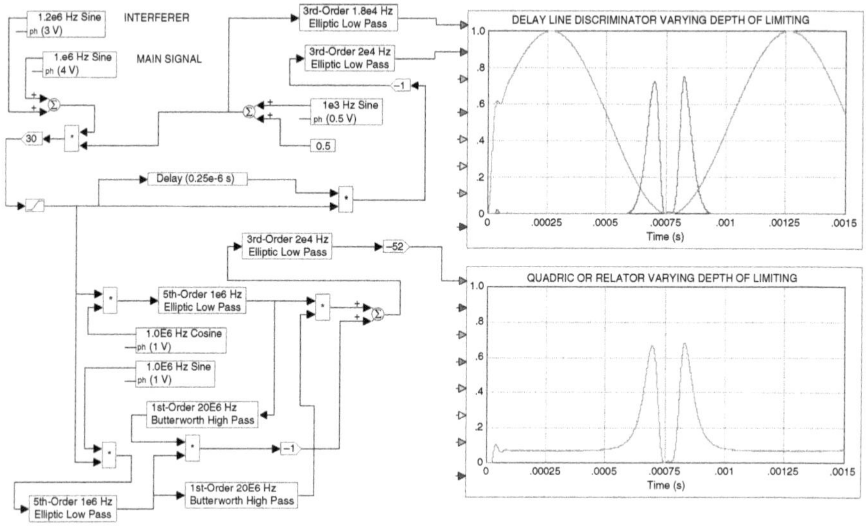

FIGURE 14.3 AM rejection does not improve appreciably when S-curve is widened between peaks. (*See the color version of this figure in color plates section.*)

Next, observe the effect of the ratio of the quadricorrelator low-pass filters' corner frequency to the interferer offset frequency, which in Figure 14.4 is now double (the interferer offset frequency is halved to 100 kHz). As expected, the quadricorrelator beat has halved in amplitude but the baseline of the quadricorrelator is now much closer to zero. Thus, we conclude that the ratio of both

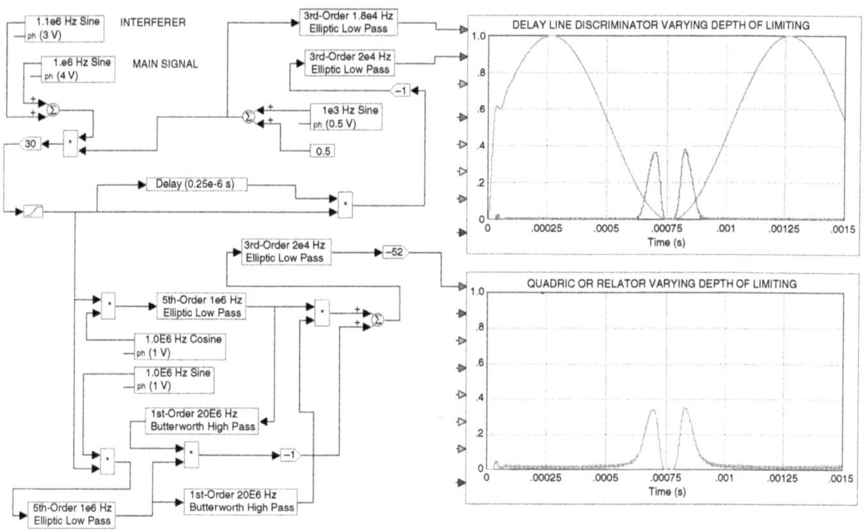

FIGURE 14.4 AM rejection improves by increasing the ratio of low-pass corners to interferer offset frequency. (*See the color version of this figure in color plates section.*)

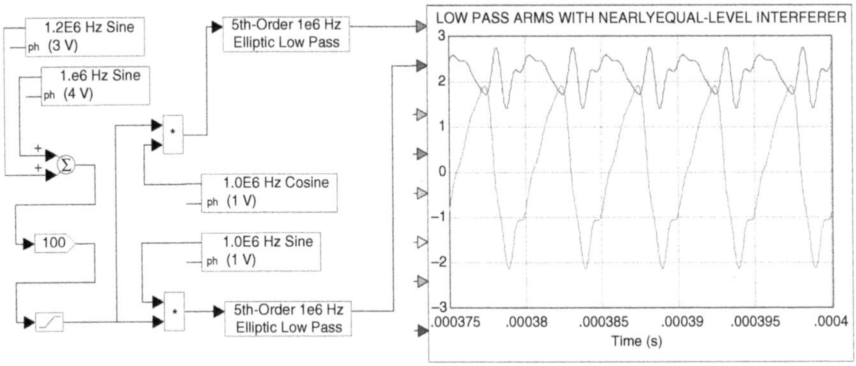

FIGURE 14.5 Quadricorrelator low-pass corner frequencies are only five times the interferer offset frequency; waveforms have kinks. (*See the color version of this figure in color plates section.*)

differentiator and low-pass corner frequencies should be at least 10 times the expected interferer offset frequency.

What is the mechanism that degrades AM rejection and interferer rejection due to the filter corners? Look at Figure 14.5 for a clue. Notice that the Q and I waveforms are replete with unexpected kinks, and these kinks disrupt the correct differentiator and subsequent multiplier outputs. Compare these kinked waveforms with the correct waveforms (having simpler kinks) of Figure 14.6, where low-pass corners are 20 times the interferer offset. Clearly, these latter waveforms are more correct due to the increase in relative filter bandwidth compared to beat frequency (i.e., more harmonics of the beat waveform can pass without truncation or delay variation). These waveforms are for illustration only since they are simulated for zero-phase angle alignment of main carrier relative to the I channel.

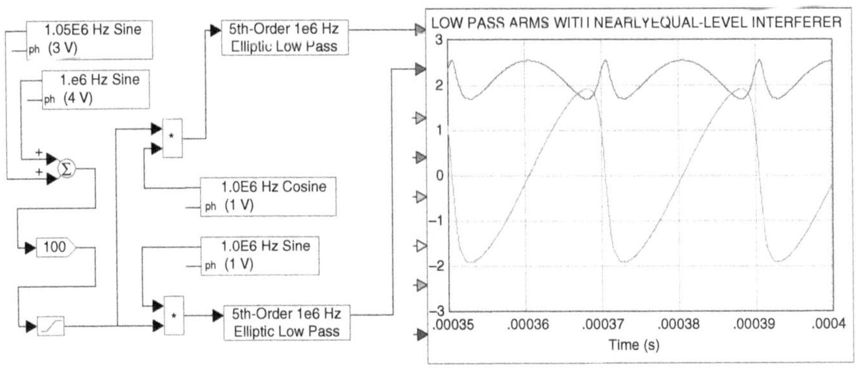

FIGURE 14.6 Quadricorrelator low-pass corner frequencies are increased to 20 times the interferer offset frequency; waveforms kinks are simpler due to greater fidelity of processing. (*See the color version of this figure in color plates section.*)

In reality, the I and Q waveforms will crawl and trade places since a slight frequency error (crawling phase) will be present corresponding to zero output voltage even in the absence of an interferer.

Notice from the AM rejection of the ideal limiter/discriminator in Figures 14.1–14.4 that the output sits very close to zero for limiter depths above about 20% of 30:1, that is, 6:1 and above. One should rerun the simulation with one's own criteria for success, namely, how "close to zero" the output actually is to the designer's objective.

14.3 IMPAIRMENT DUE TO FILTER RESPONSE TILT (ASYMMETRY)

At very high carrier frequencies, the hard limiter preceding the frequency discriminator may need an output tank circuit to resonate out any stray capacitance.

The tank circuit may be single tuned or double tuned. It will then represent a bandpass filter at the limiter output.

Observe Figure 14.7. A frequency discriminator operating at 100 MHz is fed a superimposed interferer 5 MHz away at 0.75 times the amplitude of the main 100 MHz carrier. These two signals then feed a very wideband bandpass filter, then a hard limiter (+/−1 mV clipping points), and finally a X10000 amplifier to get up to reasonable levels. The RF envelope is displayed on the right-hand plot and the demodulated beat waveform on the left-hand plot. Notice that insignificant AM

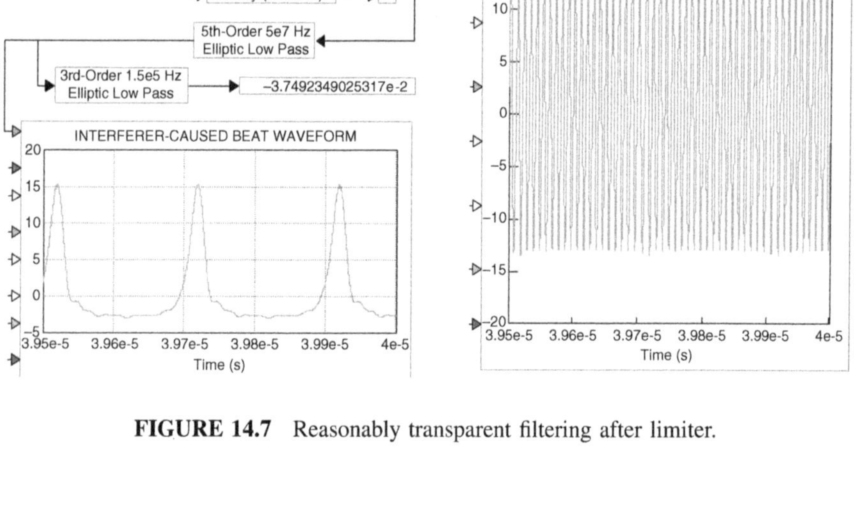

FIGURE 14.7 Reasonably transparent filtering after limiter.

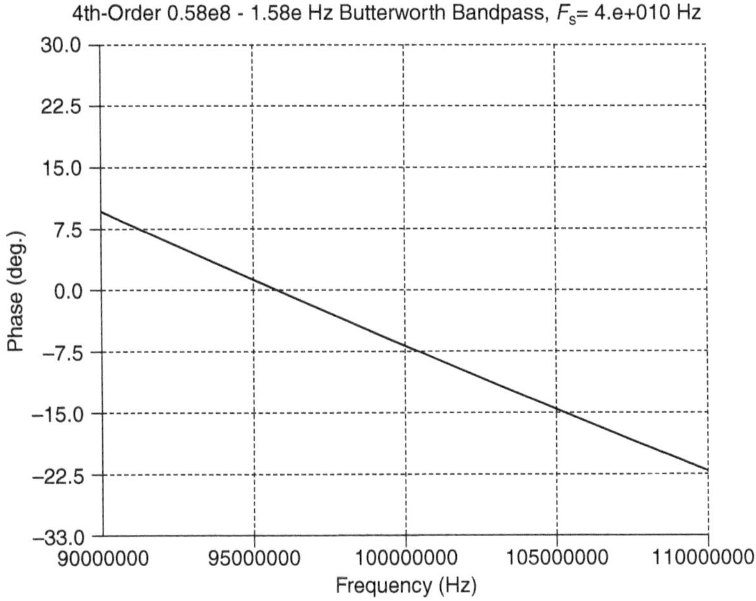

FIGURE 14.8 Filter response shows essential transparency.

envelope occurs. This is because the bandpass filter following the limiter is essentially transparent to both signals.

The voltmeter reading the average DC from the demodulator shows −37.5 mV DC, which is quite small (but nonzero due to some effect of the filter). The bandpass filter magnitude response is essentially flat for all sidebands at +/−5 MHz offset, and the phase response is nearly a straight line (Fig. 14.8), and this linear phase property is important in angle-modulation systems.

Next, we force the postlimiter bandpass filter to be asymmetrical to the sidebands +/−5 MHz around the main carrier, as in Figure 14.9. Notice that parasitic AM is now impressed upon the RF signal (right-hand plot) and our voltmeter now reads +228 mV DC. (Not so evident from the figure is the parasitic PM added by the filter.) The magnitude response of the asymmetrical filter is shown in Figure 14.10. Notice that the lower 5 MHz sideband is unmodified but the upper 5 MHz sideband is down about 1.5 dB. Figure 14.11 shows the phase response, which is no longer highly linear, as judged by the derivative of phase response (group delay) of Figure 14.12, which shows a 1 ns disparity between the delays of the lower and upper 5 MHz sidebands. The asymmetry of magnitude (and of phase, although small in this example) naturally occurs in most filters that have geometric rather than arithmetic symmetry, although exaggerated in this example. We shall define these terms soon, and explain their significance and remedy.

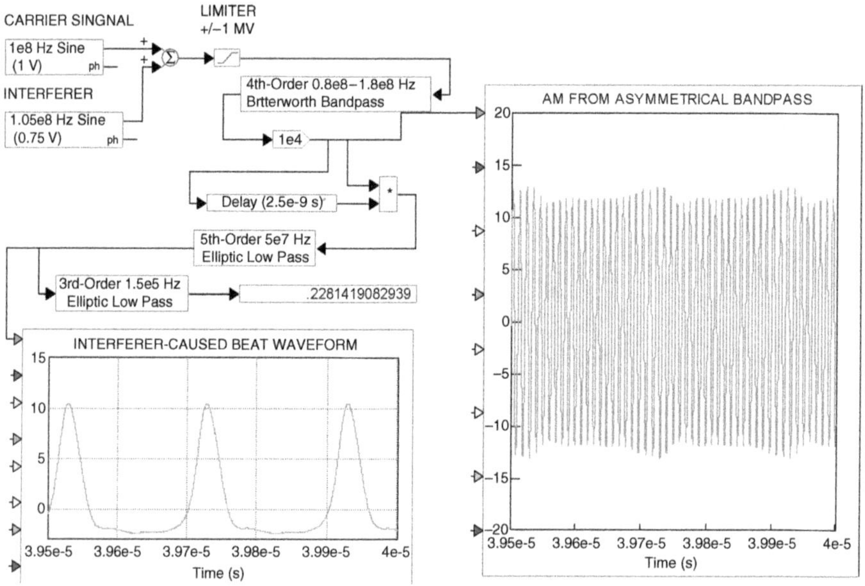

FIGURE 14.9 DC offset due to interferer and filter asymmetry after limiter.

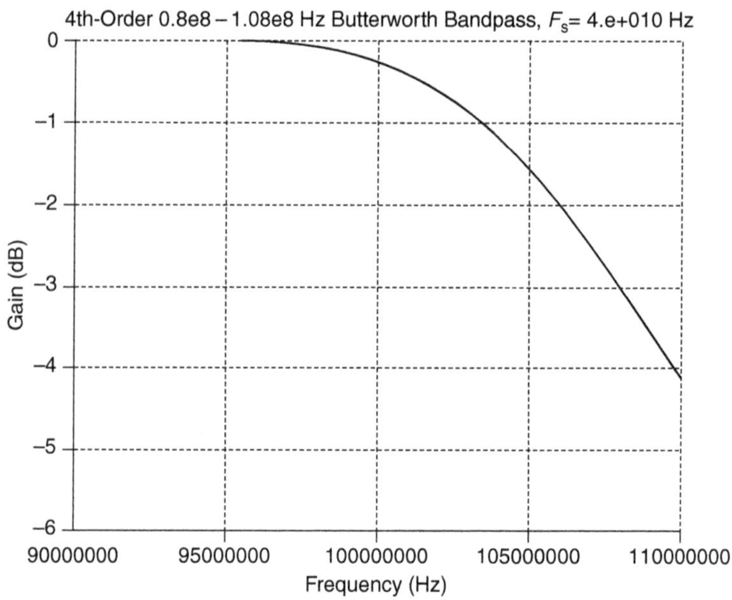

FIGURE 14.10 Asymmetrical filter magnitude response.

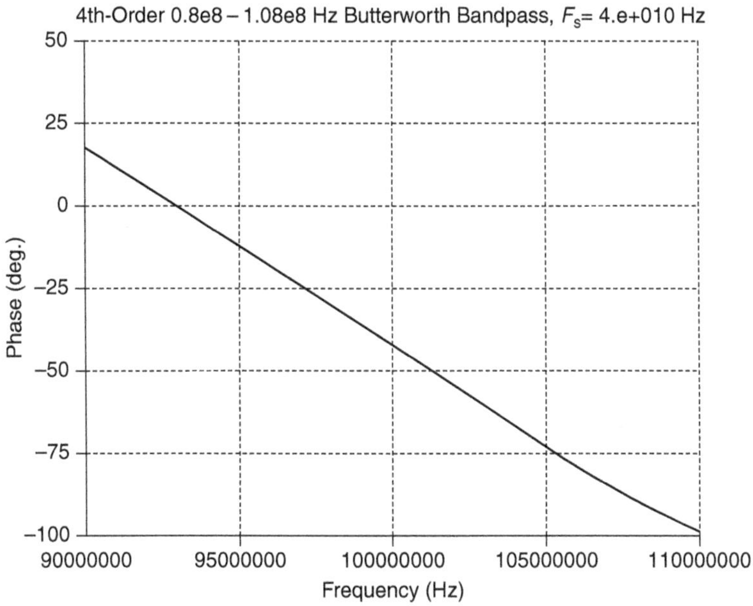

FIGURE 14.11 Asymmetrical filter phase response.

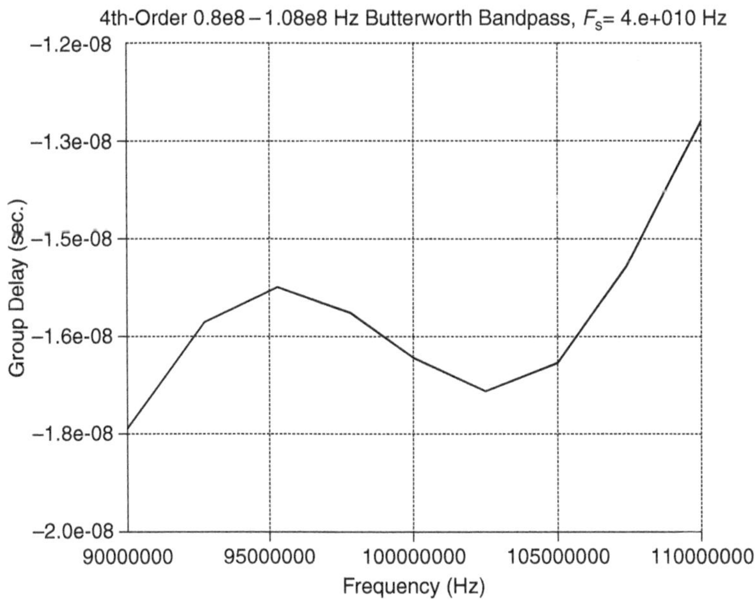

FIGURE 14.12 Asymmetrical filter group delay response.

14.4 BANDPASS FILTER GEOMETRIC AND ARITHMETIC SYMMETRY

If the postlimiter filter exists, it may be unintentionally asymmetrical by virtue of mistuning. All passive filters in general must be tuned. This may take the form of specifying tight-tolerance components, or take the form of manual tweaking. Seldom will the filter designer be so fortunate as to find off-the-shelf values for the components. Usually, the approach taken is to procure 2% tolerance capacitors, then variable inductors, since the inductors will in general not be available at precisely the value required.

But it seems valuable at this point to clarify the reason for filter asymmetry due to the type of transform employed to convert a low-pass filter to a bandpass filter. We start with a low-pass filter example of Figure 14.13. We design it for a −3 dB corner of 10 MHz and make it 12 dB/octave roll-off. If such a filter is designed to be fed by

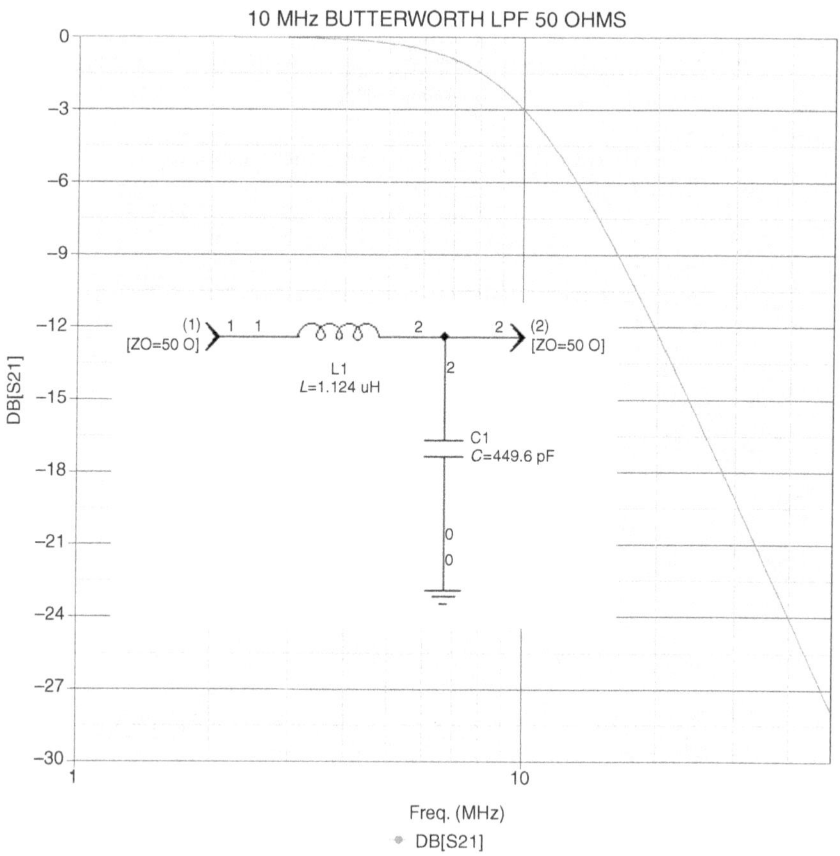

FIGURE 14.13 Magnitude response versus frequency of 10 MHz low-pass filter and topology.

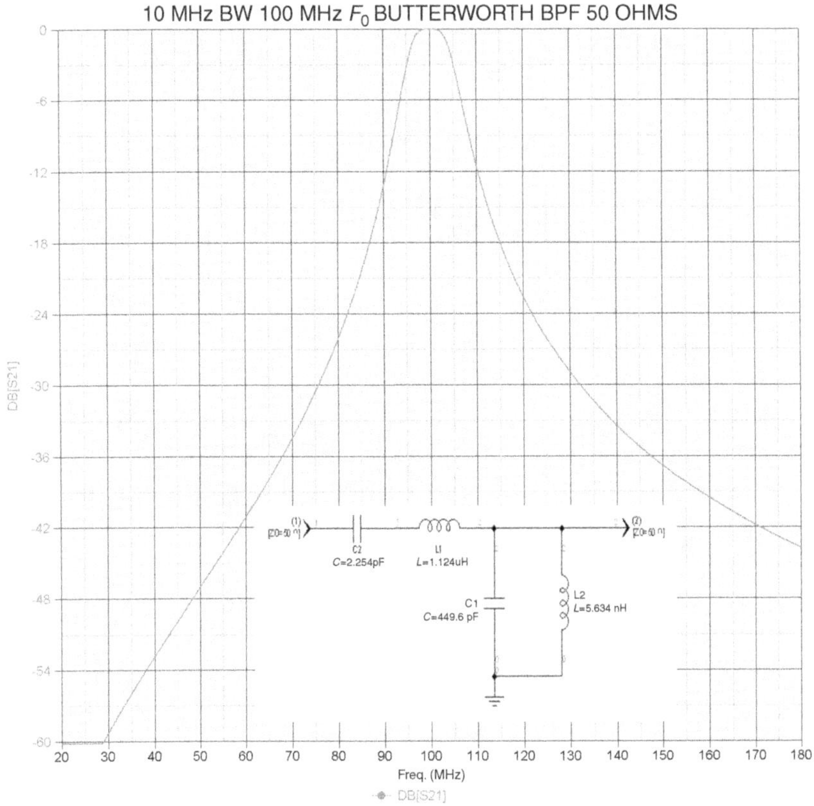

FIGURE 14.14 Magnitude response versus frequency of same 10 MHz low-pass filter transformed to bandpass and topology (note asymmetry of high side and low side stopbands).

and loaded by 50 ohms, the component values will be as shown. Next, we desire to convert this filter to a bandpass type centered at 100 MHz. The standard transform process requires that each inductor be resonated at 100 MHz with a series capacitor and each low-pass filter's capacitor be parallel resonated at 100 MHz with an inductor, to yield the filter shown in Figure 14.14, with its frequency response now as shown, centered at 100 MHz and having −3 dB points at 95 and 105 MHz (+/−5 MHz around 100 MHz). The component values that result are ridiculous for 100 MHz because L2 is too small to have high Q and L1 is too large to have negligible parasitic shunt capacitance at 100 MHz, and because C1 is too large to have negligible series inductance when connected by a reasonably short printed circuit trace. In other words, either Q issues or self-resonant issues arise. To address these realizability issues, the filter is restructured as in Figure 14.15. Notice that except for inductors, some of the capacitors aren't standard values, requiring that they be procured as trimcaps. The inductors may in general suffer the same need for adjustability for the same reason. But in spite of our trouble, the filter is even more

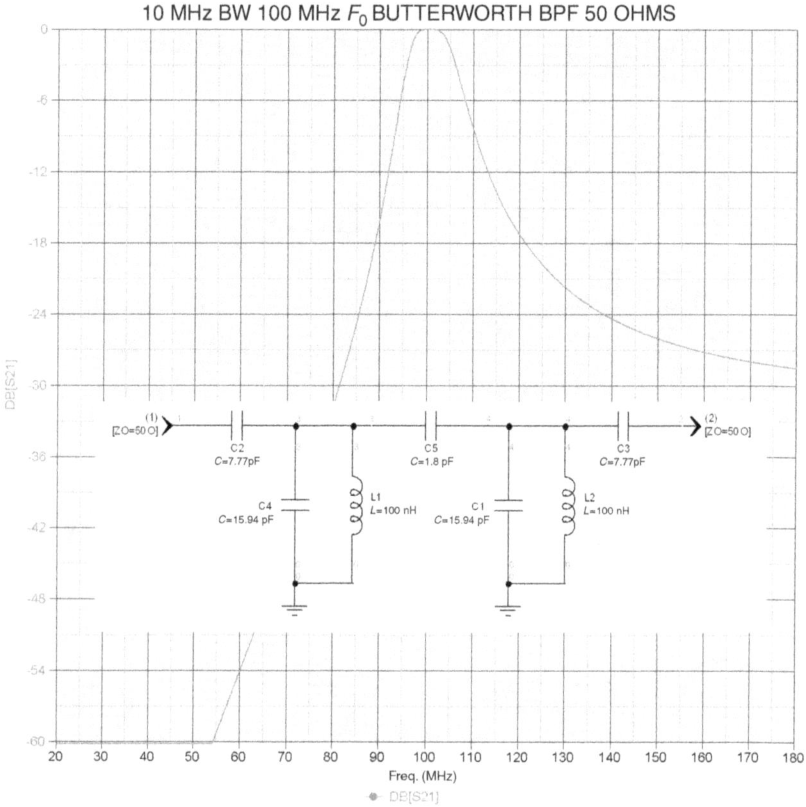

FIGURE 14.15 Coupled resonator bandpass filter response and topology using capacitive matching to 50 ohms (note severe asymmetry of high side and low side stopbands).

asymmetrical than it was (observe the high-side skirt versus low-side skirt). The filter is now called a "coupled resonator" bandpass filter due to the capacitor coupling between the parallel resonant sections. Figure 14.14 suffers from geometric symmetry, which means that response above center is not the mirror image of the response below center, and the center is the geometric mean of any two frequencies at the same amplitude. For example, if the lower -3 dB frequency is 90 MHz and the upper -3 dB frequency is 111.111 MHz, the geometric mean is exactly 100 MHz (f_0 (MHz) $= \sqrt{90 \times 111.111} = 100$), but the two -3 dB points aren't symmetrically located from the center of 100 MHz. But a filter with arithmetic symmetry might have the lower -3 dB point at 90 MHz and therefore the upper -3 dB point would be at 110 MHz.

If we alter the coupled resonator filter of Figure 14.15 so that instead of capacitive coupling between source and load we use inductors, the goal of arithmetic symmetry is clearly approximated, as in Figure 14.16. The rule of thumb in designing a globally symmetrical coupled resonator filter is to use plenty of inductors. This conflicts with the goals of cost and tuning, since inductors traditionally cost more than capacitors

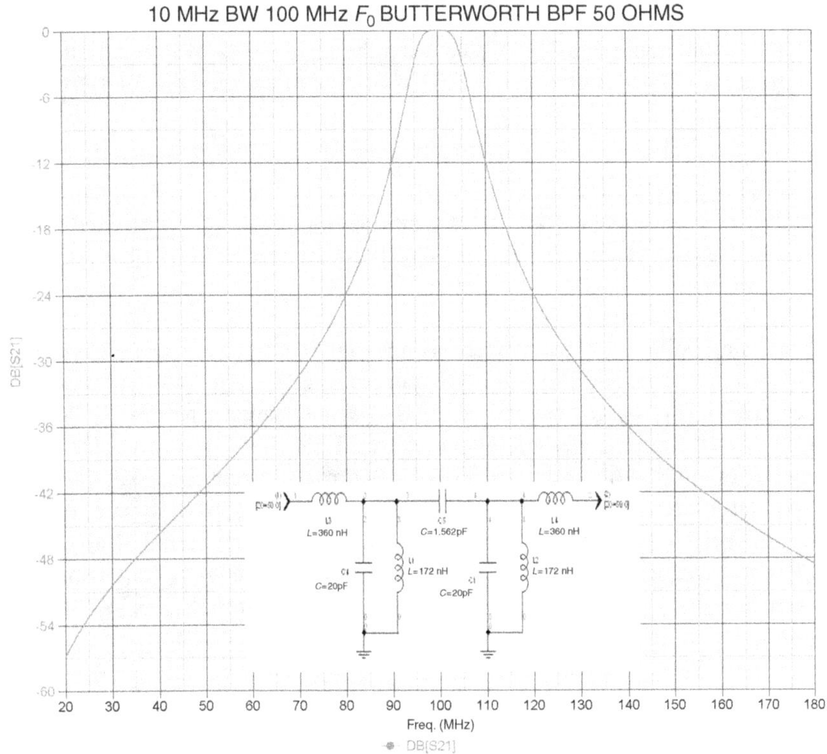

FIGURE 14.16 Greatly improved coupled resonator bandpass filter symmetry using inductive matching to 50 ohms.

and are often not available close to the design center values desired, necessitating making them variable and tweaking them ("alignment" of the filter).

14.5 COMMENTS ON DEGREE OF SCRUTINY

So far, we have scrutinized frequency discriminator performance in the presence of a very large amplitude spurious interferer. In reality, most applications will have smaller spurs, perhaps so small as to be insignificant. In such latter cases, a limiter preceding the frequency discriminator may be required only to maintain constant amplitude over frequency, temperature, and component drift. However, it has been valuable to "go for broke" in our treatment for the following reasons:

(a) Even small influences on DC output baseline of the frequency discriminator may represent many Hz or kHz of frequency error.
(b) DC baseline offset becomes more important in applications where PLL capture range is small.

(c) We have shown the quadricorrelator type of frequency discriminator to be more vulnerable to an interferer when the bandwidths of internal quadricorrelator filters are not wider by orders of magnitude than the interferer offset frequency.

(d) We have shown that filter asymmetry postlimiting affects frequency discriminator DC baseline.

(e) We have commented that beat waveform clipping at the discriminator output causes nonzero DC baseline shift due to finite headroom presented to the beat waveform.

Concerning point (e), if an interferer is not significantly present, there will be no beat waveform if a balanced quadricorrelator is used. If a large interferer is present, the beat waveform cusps can be very large. They can be greatly reduced by low-pass filtering at the differencer current output before amplification. For example, the final differencer of the quadricorrelator can output a current (current-mode summing from the two multipliers), feeding an op-amp with capacitive feedback in parallel with a resistance to convert current to voltage. That corner frequency should be much lower than the beat frequency. Similarly, the delay-line discriminator's output multiplier can be a current-mode output type feeding a similar transimpedance. We have also shown the effect of depth of limiting prior to a frequency discriminator.

15

INTERFACING THE FREQUENCY DISCRIMINATOR TO THE PLL

15.1 CONTINUOUS CONNECTION: PROS AND CONS

The quadricorrelator is probably the most flexible form of frequency discriminator because its center frequency is programmable via the reference frequency. Its output is essentially zero when there is zero frequency error regardless of phase error, so it can theoretically remain connected to the PLL all the time. When it is, it can serve the damping function. In addition, one of the multipliers in the quadricorrelator can be used as the phase detector for the PLL. This is done when the quadricorrelator operates at the same reference frequency as the PLL. Occasionally, it is preferable to divide the frequency of both VCO and reference by the same factor so that the quadricorrelator can exploit low-frequency multipliers exclusively, which may be too noisy to serve in phase detector mode for the PLL, or because phase detection is desired to happen at a high frequency to avoid the 20 log N phase noise multiplier. Another advantage of frequency division prior to the quadricorrelator is that since digital counters can be employed it is trivial to generate the quadrature VCO signals.

If the carrier is not pulsed, digital counters may be used at the VCO output and the discriminator input, but of course a pulsed frequency reference precludes standard counters. For example, a 1 GHz source and the 1 GHz VCO can

Frequency Acquisition Techniques for Phase Locked Loops, Daniel B. Talbot.

Published 2012 by John Wiley & Sons, Inc.

both be divided by 256 (resulting in a frequency comparison at just 3.9 MHz) using off-the-shelf prescalers. Again, this technique won't work if one of the signals is pulsed.

One caution should be observed. In low-pass filtering the output of the quadricorrelator, there is a parasitic pole created that if not accounted for will destabilize the frequency converging loop. However, the PLL loop integrator already contains a lead (the damping) term, so the parasitic pole is partially corrected.

15.2 CONNECTION TO PLL VIA A DEAD BAND

Refer to the VisSim folder in the toolkit and file named "quadricorr_assisted_30KHzBW_PLL_sim.vsm" corresponding to Figure 15.1. A balanced quadricorrelator drives a VCO feedback loop via a summer feeding the PLL loop integrator. The other input of the summer is fed by the low-pass filter following the I or Q multiplier, whichever we prefer, which will become the phase detector for the PLL function. Notice that a 2 V deadband is used that automatically "disconnects" the quadricorrelator's output when the loop nears phase locking. Notice that we have chosen the deadband and the gain preceding it so that the frequency assistance signal disappears as soon as the overall feedback loop converges the VCO frequency to within maybe 10 kHz of final value (which is 10.2 MHz; vertical scale for VCO control voltage is 2 V equaling 200 kHz offset from VCO resting frequency of 10 MHz). We urge the reader to tweak the value of the deadband to see its effect, as well as the gain prior to deadband. Lower gain will

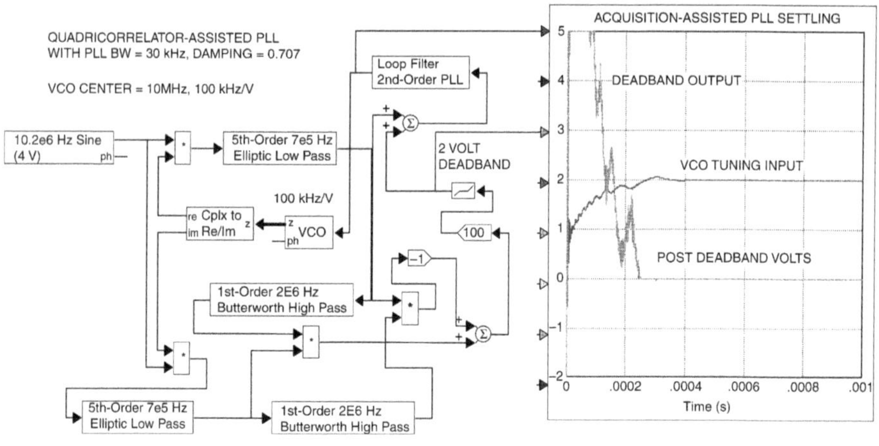

FIGURE 15.1 Simulation of quadricorrelator-assisted PLL with automatic deadband disconnect of quadricorrelator upon sufficient frequency convergence, enabling handoff to PLL without injecting noise.

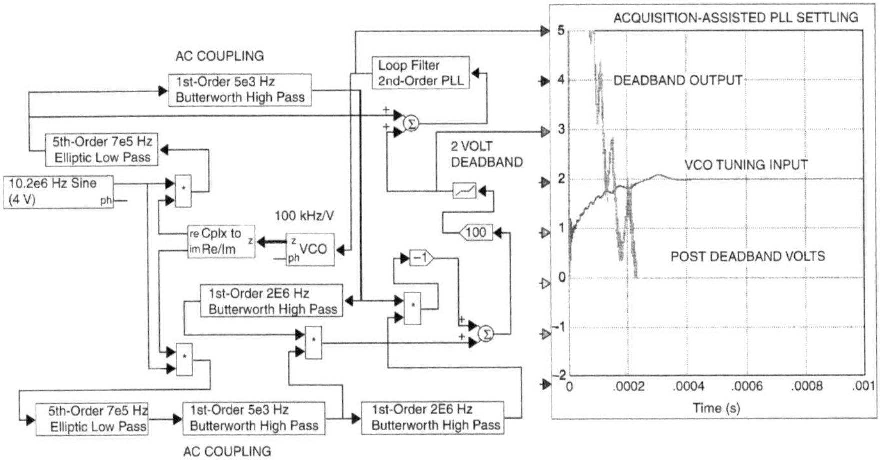

FIGURE 15.2 AC coupling the outputs of the I and Q multipliers.

cause the VCO to settle more slowly. As gain approaches zero prior to the deadband, the PLL becomes the only loop, and takes an enormous time to converge to final value. No interferer is assumed to be present, thus no limiter is needed for FM capture by the quadricorrelator. The deadband is easily accomplished via a pair of back-to-back diodes in parallel, the pair connected in series with the quadricorrelator output.

The first multipliers (mixers) in the quadricorrelator generate the I and Q signals, and in general will have DC offset at their outputs. This DC offset can be eliminated by suitable AC coupling. It is important to remember that below the cutoff of these AC coupling networks, the effect is similar to the deadband. For example, if we want to frequency converge to within 30 kHz before handoff to the PLL phase locking mechanism, we must AC couple with a corner frequency much lower than that. See Figure 15.2. Here, we used an AC coupling corner frequency of 5 kHz, and the settling time for the overall system is essentially unchanged from that in Figure 15.1 without the AC coupling. The reader is encouraged to make this modification (or run the toolkit file named "quadricorr_assisted_AC_coupled_30KHzBW_PLL_sim.vsm" in the VisSim folder) and change the AC coupling corner to 50 kHz. Run the new simulation with that value and observe a greatly modified settling time and shape.

15.3 SWITCHED CONNECTION

Observe Figure 15.3. The deadband created by the back-to-back diodes is not perfect; the diodes have capacitance, and this provides a path for high-frequency

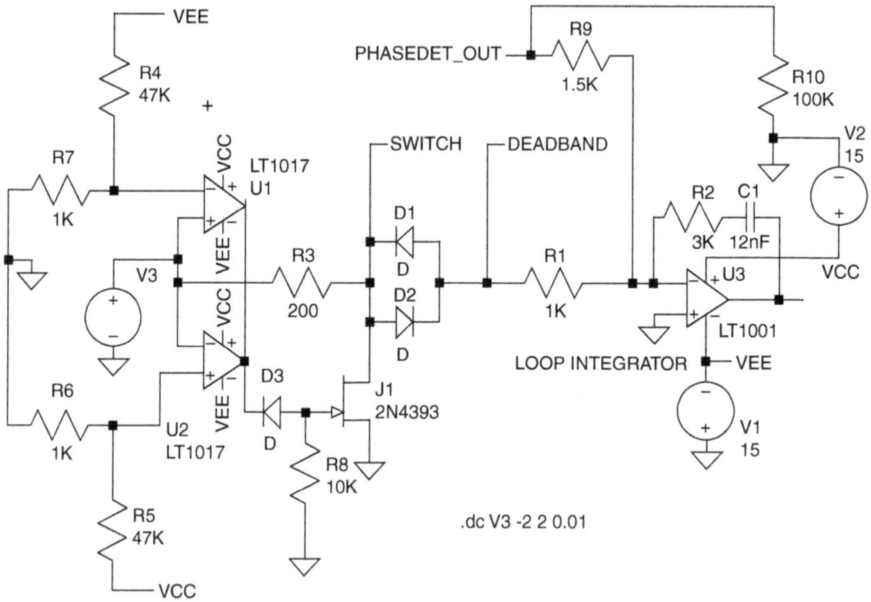

FIGURE 15.3 Schematic diagram (LTspice) of switch and deadband automatic disconnect circuitry for frequency locking.

noise to enter the loop integrator. A switch may be added to ground the signal prior to the diodes (D1 and D2) whenever it is within the deadband. The switch (JFET J1) is ON for a voltage window slightly smaller than the deadband. The drive signal for the switch is supplied by a window comparator (U1 and U2, part of a dual comparator). The deadband output is applied to the PLL loop integrator via R1. The phase detector output is applied via R9. Resistor R2 sets damping, and resistors R3 plus R1 multiplied by C1 sets the reciprocal of the integrator gain coefficient. Op-amp U3 can be as shown for most moderate noise applications, or it can be chosen for a lower noise contribution. One issue must be remembered: the noise of a 1000 ohm resistor at room temperature is 4 nV per root Hz. This is larger than the input noise of some newer "low-noise" op-amps. Therefore, the designer is urged to select impedances and op-amp current specifications carefully.

Figure 15.4 shows the resulting DC sweep transfer function of the deadband-plus-switch circuit of Figure 15.3. The waveform with largest amplitude is the window comparator output scaled by 0.2 to make it more comparable to the other waveform scales (the actual waveform swings from rail to rail or -15 to $+15$ V). The deadband output transfer function is shown having the widest window horizontally, and the prediode DC transfer function is shown having a step both before and after the window comparator pulse (looking left to right).

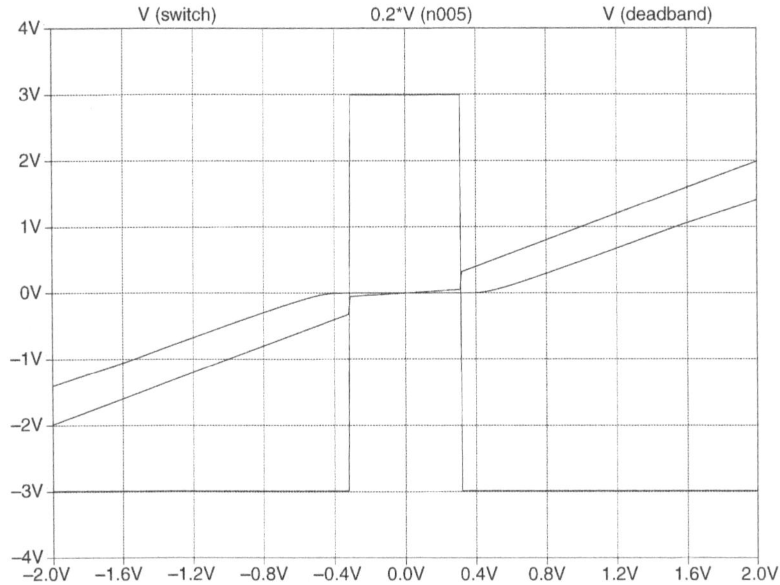

FIGURE 15.4 DC transfer function pre- and postdeadband when switch is present.

Figure 15.5 shows the revised schematic diagram with the window comparator and switch circuitry removed, leaving just the deadband diodes and associated circuitry. Figure 15.6 shows the DC transfer function.

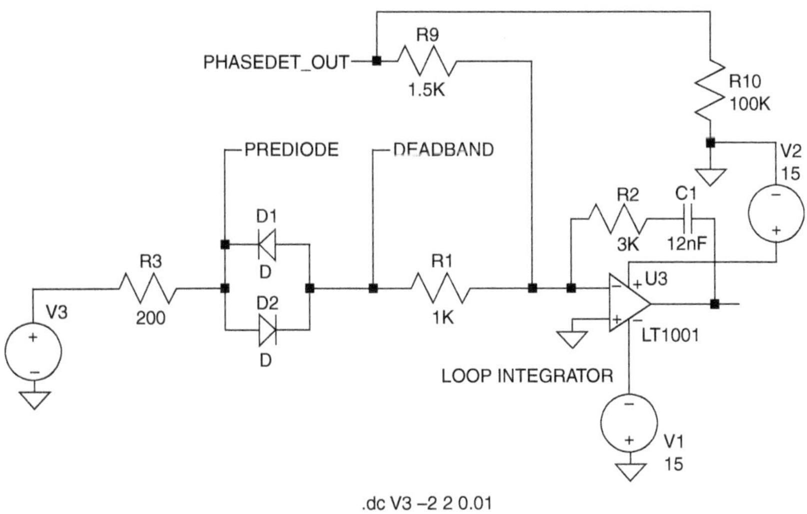

FIGURE 15.5 Automatic frequency locking deadband disconnect only.

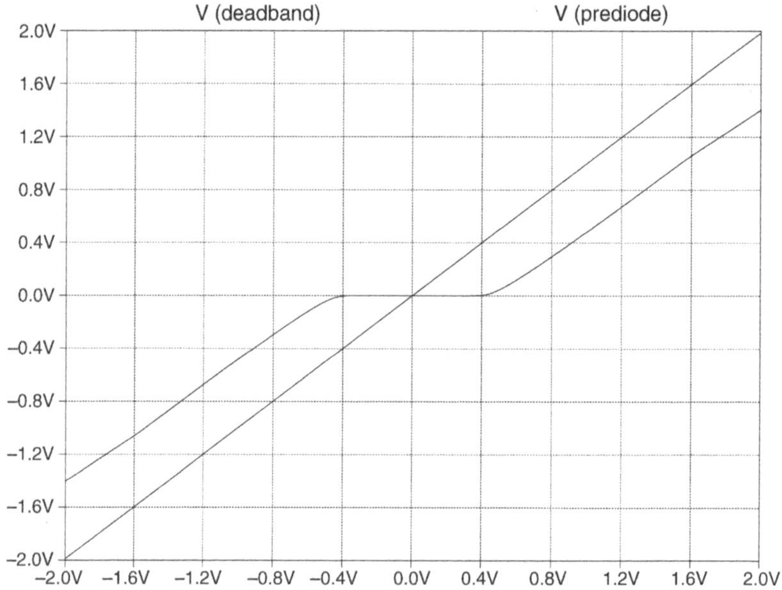

FIGURE 15.6 DC transfer function of deadband-only circuit of Figure 15.5.

16

ACTUAL FREQUENCY DISCRIMINATOR IMPLEMENTATIONS

16.1 QUADRICORRELATOR, LOW-FREQUENCY IMPLEMENTATION

Figure 16.1 shows an actual working circuit implementation of a quadricorrelator operating at 55–90 MHz. Mixers U34 and U35 decompose the reference frequency input signal into I and Q baseband components. Quadrature splitter U42 accepts a +12 dBm (about 16 mW RMS) waveform from the buffered VCO output and derives zero and 90° copies at about +7.5 dBm each. An in-phase splitter is formed by R77, R78, and R79 to drive the RF mixer ports at about 6 dBm each (this is a resistive splitter having 6 dB loss to each port, so the reference signal needs to be 12 dBm. Since these amplitudes drive the mixer into compression, limiting is automatically accomplished without an explicit separate limiter circuit.

The I and Q signals exit the IF port of the respective mixers, which are then filtered by third-order Cauer filters (C94, C96, C98, and L11 for the I signal; C95, C97, C99, and L12 for the Q signal) and then AC coupled to the next stages via C100 and C101 driving R90 and R91, respectively. These signals are then buffered by U27B and U27A and their associated feedback resistors R32 and R33.

The buffered and filtered I and Q signals are processed by the multiplying and differentiating circuitry of Figure 16.2 . U37, R102, C103, and R35 comprise the differentiator for the Q channel; U38, R103, C104, and R36 comprise the differentiator for the I channel. These derivatives then each feed an analog

Frequency Acquisition Techniques for Phase Locked Loops, Daniel B. Talbot.

Published 2012 by John Wiley & Sons, Inc.

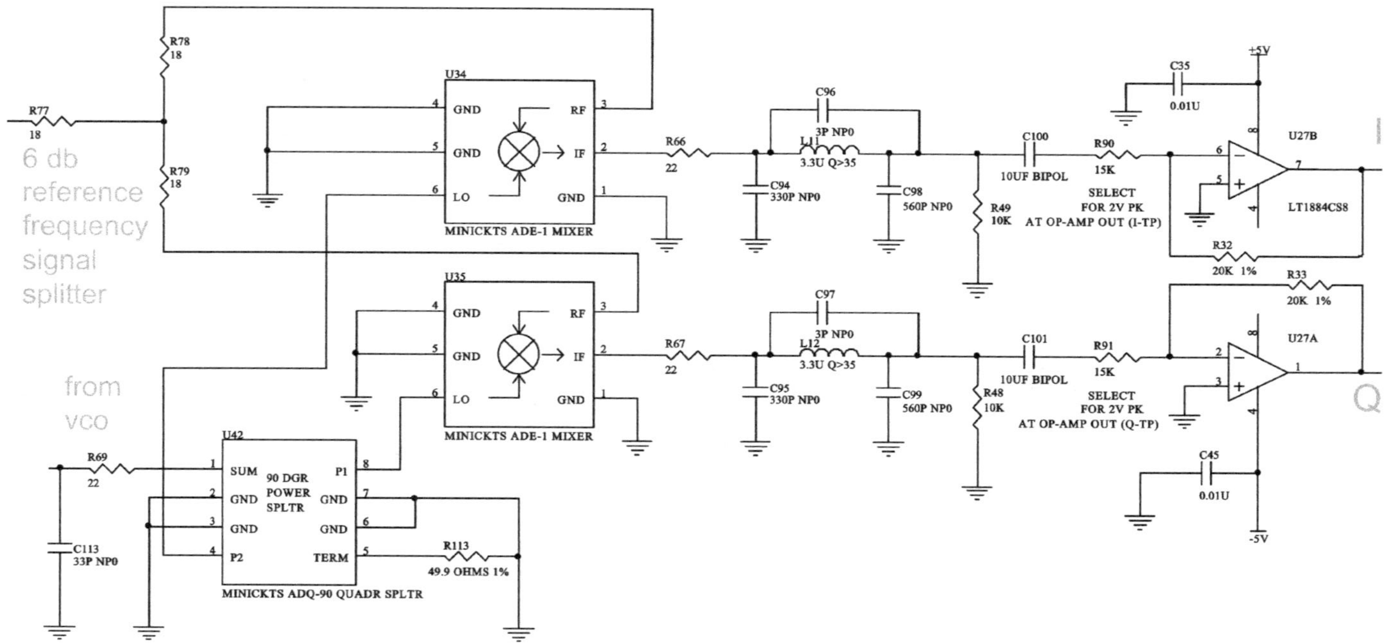

FIGURE 16.1 Schematic diagram of quadricorrelator I and Q decomposing circuit implementation.

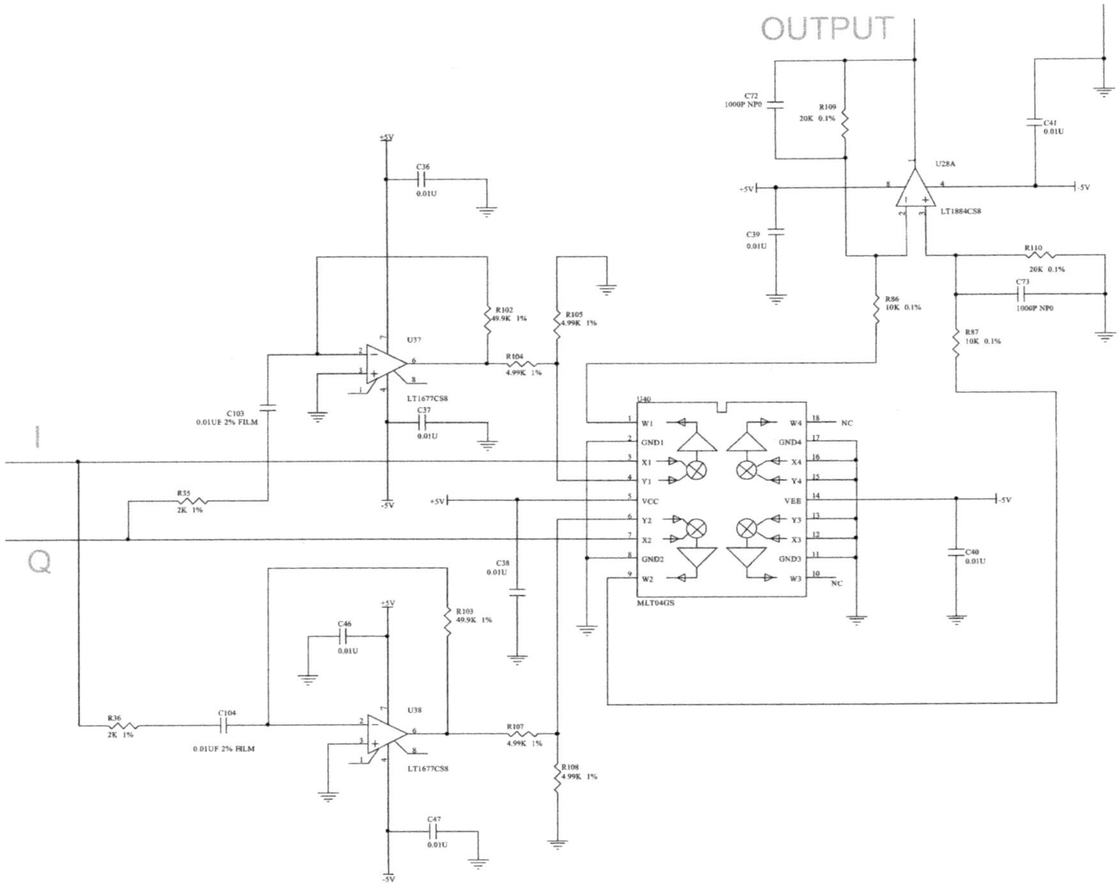

FIGURE 16.2 Remaining circuitry schematic diagram of an actual implementation of quadricorrelator.

multiplier (part of U40) along with their nonderivative Q and I baseband signals. The analog multiplier outputs are then differenced and amplified by op-amp U28A and its associated resistors.

The phase error signal for the PLL function could have been taken from the IF port of either mixer U34 or U35 (of Fig. 16.1), but the decision was made to use a separate phase detector for a variety of reasons.

16.2 FREQUENCY RATIO CALCULATING CIRCUIT FOR WIDE-BANDWIDTH USE

For extremely wide frequency range applications, the frequency ratio measuring circuit of Figure 16.3 may be useful. Its advantage is that it uses no capacitors or inductors so it can be completely realized in an IC (integrated circuit). The final stage is an R-2R-4R (etc.) resistive-summing DAC (digital to analog converter) without need for a voltage reference (it uses the 5 V logic supply and depends on it for final accuracy). Its output rests at exactly half the 5 V logic supply voltage when the two frequencies (reference and VCO) are equal (ratio = 1). Figure 16.4 shows the transfer function, with ratio of VCO to reference frequency plotted on the x-axis and output voltage on the y-axis. All resistors are 0.1% tolerance types, preferably thin-film unless temperature coefficient is not an issue.

The basic architecture is that of the classical frequency counter, where the number of VCO pulses occurring after 128 reference frequency pulses are counted and output as a voltage rather than a displayed number. The counters are 8-bit resettable upcounters.

The final logic device is a latch that holds the value of the last count and ignores the reset command of the previous counter. The circuit as shown is capable of working at frequencies under 5 MHz and ratios of VCO to reference frequency of 0.2–2. If the resistors are perfect, the output rests at half the VCC rail within about 1% for a unity frequency ratio. Accuracy and resolution can be improved by using counters and latch having more bits.

Figure 16.5 shows the block diagram for the frequency ratio detector. Pulses coming from the VCO divided by M ($M \geq 1$) are fed to an AND gate, to be counted by an N bit upcounter until the counter is cleared by the MSB-bar pulse from a separate N counter clocked by the reference counter fed by reference frequency over M. This results in 127 counts from the reference counter if it is 8 bits wide. This allows for the VCO/M counter to count up to its maximum before overflow, thus able to count a number of VCO/M pulses equal to twice the number of counts up to the MSB of the reference counter. Thus, the expression for maximum permissible frequency ratio for the example is VCO/$M \geq 2*$REF/M.

Before clearing the VCO counter, we want to read and store its output via a latch (digital sample and hold), which is simply a D flip-flop array. Then, after a "small delay" (provided by a cascade of hex inverters and logic elements), the counter can be cleared. Its output value is stored and output by the latch. Its digital value is converted to analog voltage (if desired) by a DAC. The process repeats, and the latch

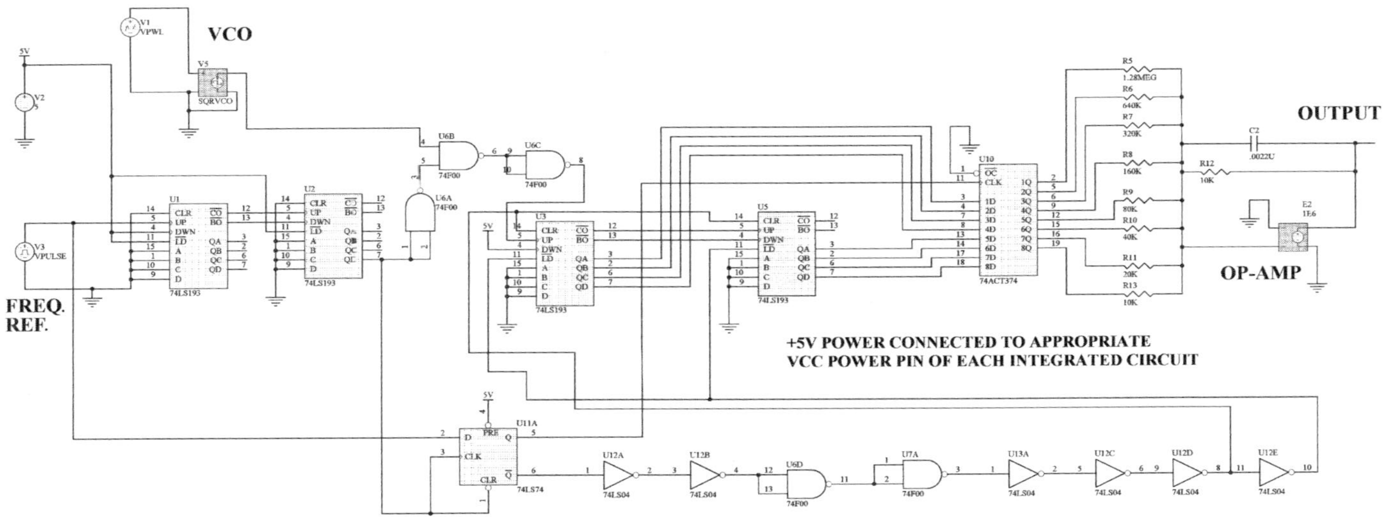

FIGURE 16.3 Actual circuit diagram of a frequency ratio detecting circuit.

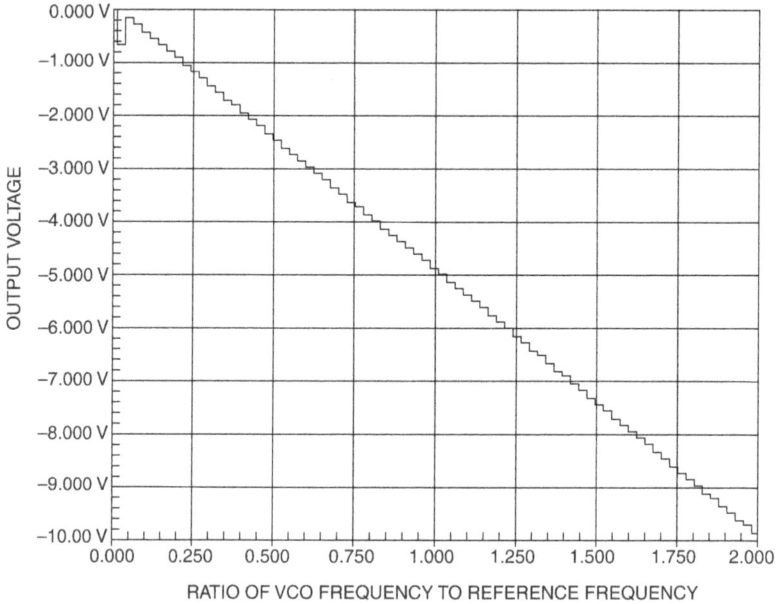

FIGURE 16.4 DC transfer function of circuit of Figure 16.3 (with ratio of VCO to reference frequency on x-axis, output voltage on y-axis).

is updated to the same value if no frequency shifts have occurred, except for a small quantization error, familiar to most frequency counter users as "display bobble." The bobble is evident in the curve of Figure 16.4. The lowest VCO/M frequency that can be counted and measured for practical purposes is the reference frequency over 2M, making this approach useful over at least two octaves (VCO/[2M] to VCO∗[2/M]). This circuit is useful at low frequencies, such as motor control, and octave-wide PLLs.

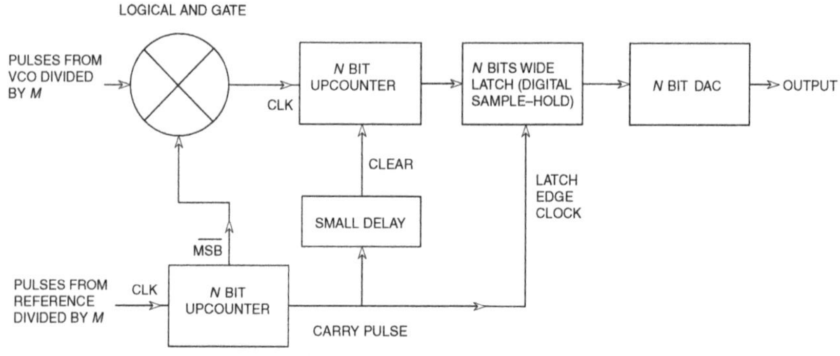

FIGURE 16.5 Block diagram of the frequency ratio detector circuit.

16.3 DIVIDING THE FREQUENCY AND RESULTANT IMPLEMENTATION

Sometimes, the frequency detector can be implemented at lower frequencies where components become cheaper and more available. In such cases, the circuitry of Figures 16.6–16.8 may also be useful. It has a narrower operating frequency excursion capability than the ratio circuit discussed previously, but has the very distinct benefit of having a zero-baseline near ground, and a bipolar output swing. In addition, it is linear in converting Hz to volts, not ratiometric.

The actual circuit values were targeted for a PLL operating at 13.333 MHz. The VCO was a 53.333 MHz quartz oscillator whose frequency was pullable and uncertain

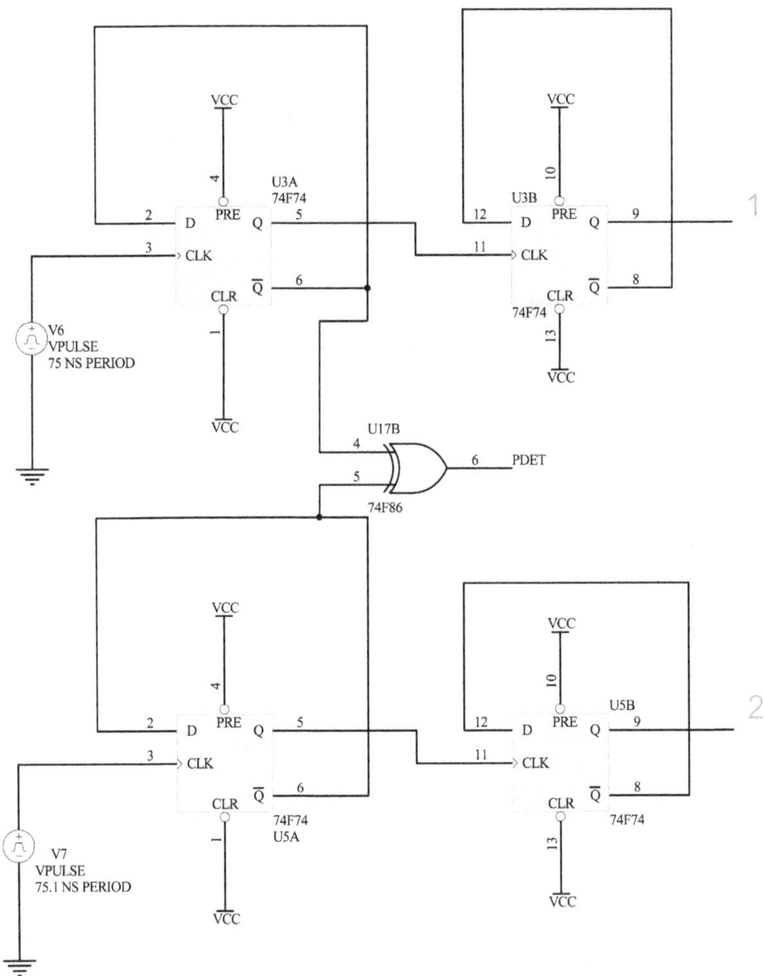

FIGURE 16.6 First stages of low-frequency operating frequency discriminator.

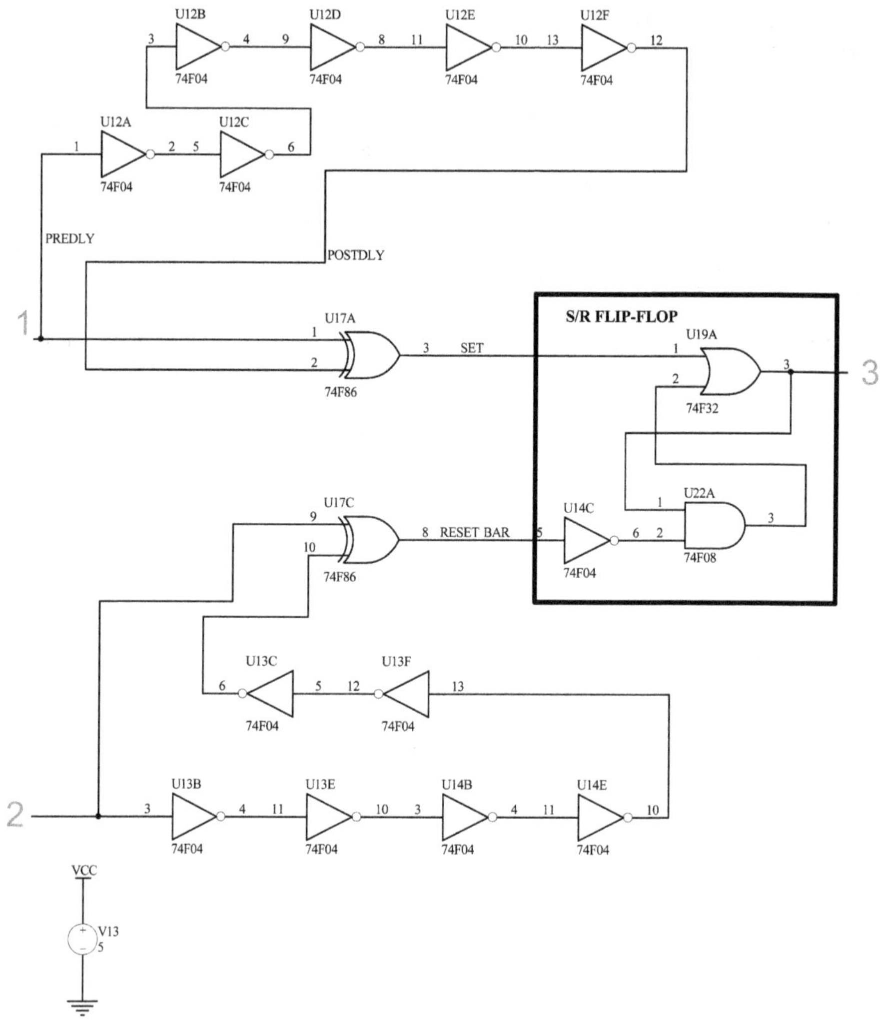

FIGURE 16.7 Second stages of low-frequency operating frequency discriminator.

by +/−3 kHz. The PLL was designed for a jitter measurement application, so the low noise of the quartz unit (quartz crystals can have Q well over 10,000, occasionally a million, so the Leeson knee is very low) was exploited. This required a PLL having low bandwidth (100 Hz) in order to add minimal noise to the VCO. Naturally, a 100 Hz PLL BW cannot easily or quickly acquire and converge from a 3 kHz offset, so frequency acquisition assistance was dictated. Hence, the circuitry to be discussed. In Figure 16.6, the reference frequency of 13.333 MHz is represented by a pulse signal of 75 ns period, V6. The VCO/4 signal (division by 4 occurred elsewhere) is represented by V7, set at 17 kHz lower than the reference frequency for assurance that the circuit will work at an even higher offset frequency than the required, and to speed simulation

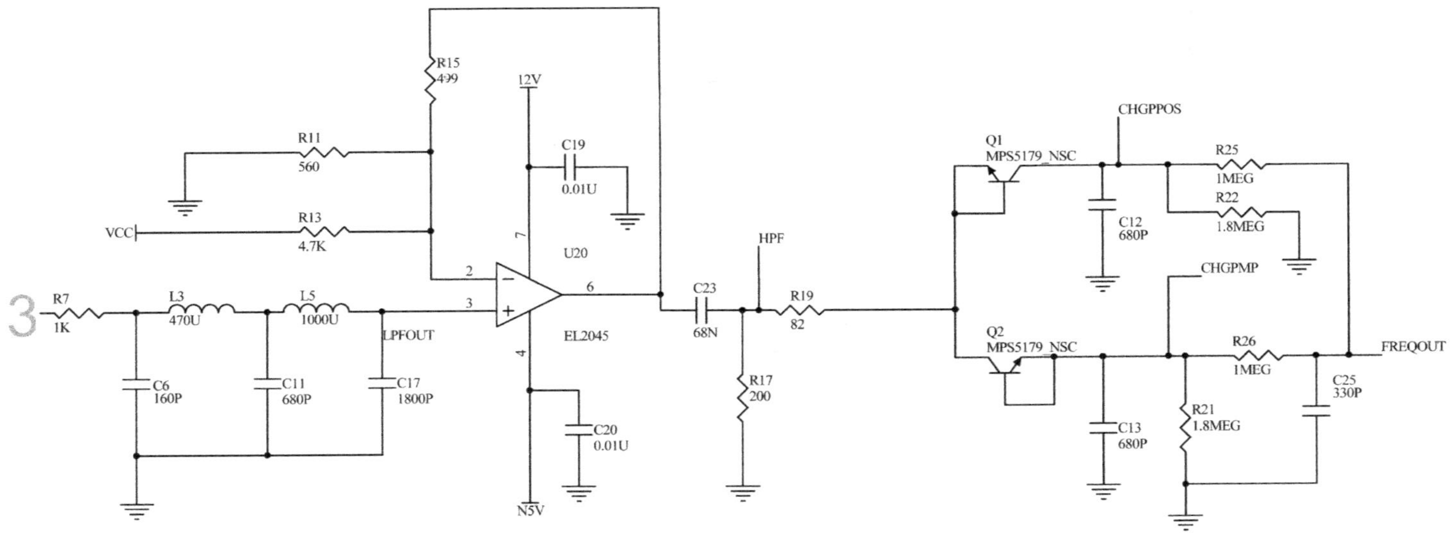

FIGURE 16.8 Final stages of low-frequency operating frequency discriminator.

by a digital-capable SPICE program. Phase detection was done by exclusive OR gate U17B at half frequency of V6, via divide-by-2 D flip-flop U3A (and U5A for the already-divided-by-4 VCO pulses). The output of the phase detector was processed by a standard type-II PLL. Of course, this type of phase detector's DC baseline is VCC/2 volts, so great effort was made to subtract that baseline as noiselessly as possible prior to the loop integrator. And low-pass filtering was employed ahead of the loop filter to remove the pulses from the ex-OR gate U17B and leave only DC duty factor driven excursions but not so much filtering to make an unstable PLL (not difficult when the PLL BW is only 100 Hz and the phase detector pulses are at 6.6666 MHz). But so much for the standard PLL operation.

On the right edge of the drawing (Fig. 16.6) are two points marked 1 and 2, which marry with the same number points on the left edge of the next drawing (Fig. 16.7). There are two more flip-flops that divide the frequency by another factor of 2 (U3B and U5B in Fig. 16.6). The signals at this point (points number 1 and 2) are square waves due to the flip-flops. The edges of these square waves are used to trigger an S/R flip-flop (U19A and U22A in Fig. 16.7). These edge triggers are very brief owing to the delay of the cascaded hex inverters (U12 and U13–U14) and their associated exclusive OR gates (U17 A and C). The output of the S/R flip-flop appears at node marked "3" in the figure, said node marrying with the next figure's (Fig. 16.8) same numbered node. The low-pass-filtered signal at node number "3" appears similar to the ramp-and-sample phase detector, which is a sawtooth wave when a frequency error exists, whose slope is positive or negative depending on the sign of the frequency error. The low-pass filter consists of R7, C6, L3, C11, L5, and C17. This sawtooth waveform is then amplified and impedance buffered by the high current capability of op-amp U20. It is then AC coupled to a diode detector network very similar to that of Figure 5.3. Spikes of current are limited in amplitude by R19 in order to protect the diodes formed by Q1 and Q2. Low-capacitance transistors are used as diodes because they afford high speed and low leakage (lower than a

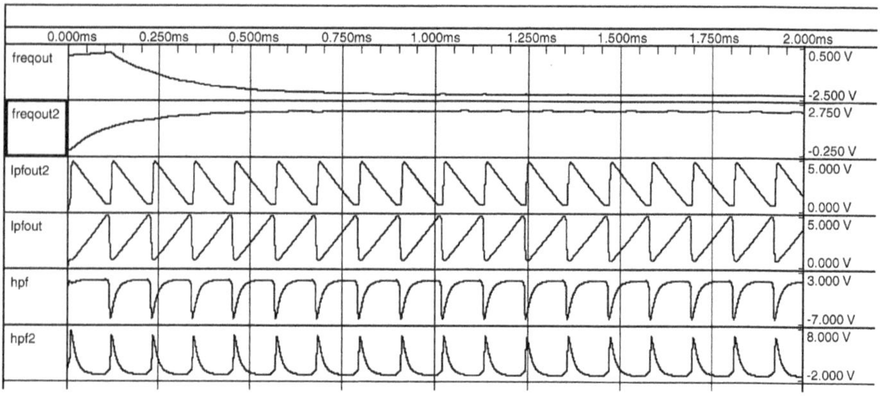

FIGURE 16.9 Waveforms of frequency discriminator with a frequency error, showing sawtooth shapes and DC output.

conventional diode such as a 1N4148). This low leakage current is required since the load to the diode network at DC is very high impedance (1.8 megaohms). The final output of the circuit is labeled "FREQOUT" on the right edge of the figure.

The waveforms corresponding to Figure 16.8 are shown in Figure 16.9 . Notice that they strongly resemble the waveforms produced by the circuit of Figure 5.3 that were discussed and pictured in detail. LPF and LPF2 are the output waveforms from the low-pass filter for two equal-frequency offset but opposite polarities; HPF and HPF2 are the corresponding differentiated waveforms applied to the diode network. Freqout and Freqout2 are the respective diode network output waveforms after filtering or averaging.

16.4 MARRIAGE OF BOTH FREQUENCY AND PHASELOCK LOOPS

Much of the discussion of interfacing the frequency acquisition assistance circuitry to the phase lock circuitry has already been covered in Chapter 15. However, in this section we will give examples of marrying the frequency discriminator (quadricorrelator) and phase loops. This will be done for the case of a deadband to disconnect the frequency acquisition assistance after phase lock, and for the case where no disconnect is employed.

Before we begin the examples, let us establish the following fundamentals. First, for a type-II second-order PLL, there is a quantitative relationship between damping factor, natural frequency, and −3 dB bandwidth. The −3 dB point can be computed from the program "F3DB-OVER-FN-VS-DAMPING.EXE" in the Handy Tools subfolder of the folder "Supplemental Software for Book" in the toolkit. The −3 dB point is plotted in Figure 16.10 for damping factors between 0.5 and 2. Second, the gain of two types of phase detectors is calculable from the plots of Figure 16.11 . If an overdriven mixer or multiplier or exclusive OR gate is used as a phase detector, it will have a triangular periodicity, and clearly the conversion gain is $2/\pi$ volts per radian multiplied by the peak voltage produced by the phase detector (which can be quickly measured in circuit by forcing an out-of-lock condition so that the amplitude of the beat waveform from the phase detector can be observed after a low-pass filter to remove carrier and harmonic components). Thus, if the triangular phase detector produces a peak voltage of 2 V, the gain is $4/\pi = 1.273$ V per radian. If, however, the phase detector's beat waveform appears sinusoidal, its gain is simply E_{peak} volts per radian. Thus, the sinusoidal phase detector's gain is $\pi/2$ times greater than the triangular phase detector having the same peak amplitude.

Load and run the program in the toolkit named "pllsynth.exe." Enter values for VCO gain of 100 kHz/volt, phase detector gain of 2, natural frequency of 30 kHz, damping of 1.0, and $N = 1$ for both minimum and maximum values. The program returns the values for loop integrator components; the input values screen looks like Figure 16.12 , the output values screen looks like Figure 16.13 . The loop integrator values are CF = 23.6 nF, RIN = 1.5 K, and RF (damping) = 450 ohms (in series with CF). The key constants are RIN and CF, whose time constant is RIN∗CF = 35.4 μs.

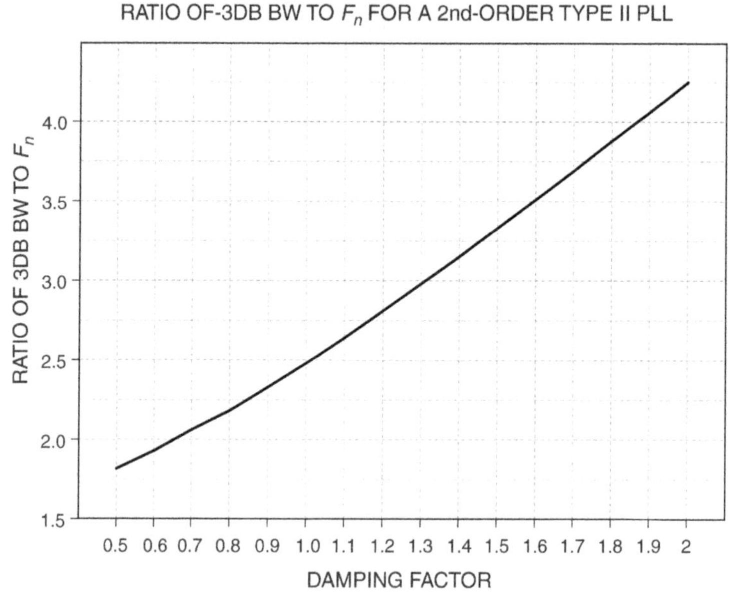

FIGURE 16.10 BW_{3dB}/F_n vs Damping in type-II 2nd order PLL.

The reciprocal of this time constant is equal to the gain of the loop integrator, or $2.825e+04\,s^{-1}$. This would be the coefficient to be used for K_i that in simulations multiplies the 1/s of the loop integrator (for block diagram simulations such as VisSim). The transfer function for the closed loop reveals that the natural frequency

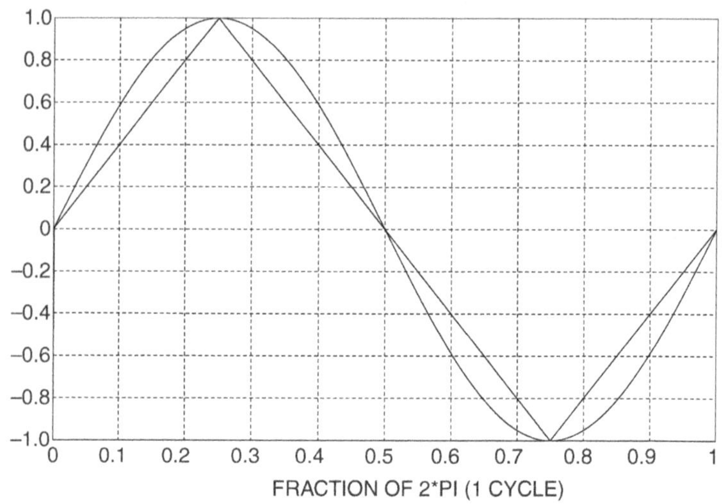

FIGURE 16.11 Sinusoidal vs. Triangular Phase Detector.

```
* * * * * * * * * * * * PROGRAM NAME = 'PLLSYNTH' * * * * * * * * * * * * *
CALCULATES PLL ACTIVE FILTER COMPONENT VALUES AND OUT-OF-BAND FM NOISE

VCO SENSITIVITY IN HERTZ PER VOLT        =? 1e5
PHASE DETECTOR GAIN IN VOLTS PER RADIAN =? 2
REQUIRED LOOP DAMPING FACTOR      D      =? 1
REQUIRED LOOP NATURAL FREQUENCY    F0    =? 3e4
SMALLEST FREQUENCY-DIVISION RATIO   N    =? 1
LARGEST  FREQUENCY-DIVISION RATIO   N    =? 1
OP-AMP INPUT VOLTAGE NOISE PER ROOT-HZ   =? 0
```

FIGURE 16.12 Pllsynth.exe input screen for example.

```
OP-AMP INPUT VOLTAGE NOISE PER ROOT-HZ   =? 0
*****************************************************************************
FOR THE CASE WHERE THE DIVISION RATIO N = 1

     CF =  2.357853E-8    FARADS

     RIN = 1.5 K   OHMS

     RF =  450    OHMS

     FM =  0 HZ RMS DEV   PER ROOT-HZ FM DET BW (EXCLUDES RESISTOR NOISE)
           WELL ABOVE THE LOOP NATURAL FREQ  F0 = 30000

PULL-IN RANGE FOR NO CYCLE-SLIPPING IS APPROX = 30000 HZ
-----------------------------------------------------------------------------
FOR THE CASE WHERE THE DIVISION RATIO N = 1

SIMPLY CHANGE RIN (OR AN EQUIV EFFECT) TO 1500  OHMS

OUT-OF-BAND NOISE FLOOR WILL NOW RISE TO 0 HZ DEV RMS PER ROOT-HZ
FM DET BW EXCLUSIVE OF RESISTOR NOISE  (WELL ABOVE F0)

PULL-IN RANGE FOR NO CYCLE-SLIPPING IS APPROX = 30000 HZ
TO END, HIT ENTER? _
```

FIGURE 16.13 Pllsynth.exe output screen for example.

f_n is proportional to the square root of this integrator gain if all other coefficients are unchanged. Thus, to scale the loop f_n to 15 kHz, we must decrease K_i by four times, making it equal to 7060. The salient calculations made by "pllsynth.exe" are as follows, in the BASIC programming language:

```
190 INPUT "VCO SENSITIVITY IN HERTZ PER VOLT =";V
210 INPUT "PHASE DETECTOR GAIN IN VOLTS PER RADIAN =";P1
220 V = V*P1: REM PRODUCT OF VCO AND PHASE DET COEFF'S
240 INPUT "REQUIRED LOOP DAMPING FACTOR D =";D
260 INPUT "REQUIRED LOOP NATURAL FREQUENCY F0 =";F0
280 INPUT "FREQUENCY-DIVISION RATIO N =";X1
***********************************************************"
360 PRINT "FOR THE CASE WHERE THE DIVISION RATIO N =";
370 PRINT X1
380 PRINT
```

```
390 TWOPI = 2*3.14159
400 CF = V/(TWOPI*F0*F0*X1*1500!)
410 REM RIN = 1500 OHMS (FOR LOW JOHNSON NOISE WITH MOST OP-AMPS)
420 RF = 2*D/(TWOPI*F0*CF)
450 PRINT " CF = ";:PRINT CF;:PRINT" FARADS"
470 PRINT " RIN =";:PRINT " 1.5 K OHMS"
490 PRINT " RF = ";:PRINT RF;:PRINT" OHMS"
```

Load and run the file named “quadricorr_assisted_20KHz_Fn_PLL_sim.vsm” in the VisSim subfolder in the SUPPLEMENTAL SOFTWARE FOR BOOK folder in the toolkit. The screen should look like Figure 16.14 . Running the simulation yields a well-damped time response from the VCO control voltage (hence frequency settling) with phase settling occurring shortly thereafter. This PLL uses the quadricorrelator to provide damping, so the integrator series feedback damping resistor is absent altogether. Of course, one could split the damping contribution into two parts, one by including a portion of the damping by series feedback resistor and a lesser portion from the quadricorrelator. The ratio of the two contributions is the designer’s choice in trading off quadricorrelator noise contribution to the loop versus frequency convergence time (and the overcoming of DC offsets). For this example, 100% of the damping is provided by the quadricorrelator. Because the integrator model has a unity gain, the gain ahead of it must be as we just computed earlier, namely, $(\text{RIN}*\text{CF})^{-1}$. We have chosen a value of 8000 to make a PLL having a natural frequency f_n slightly less than 20 kHz and we chose a quadricorrelator strength of 700,000. Let’s “kick the tires” a bit. Manually edit the value of the quadricorrelator strength from 700e + 03 to

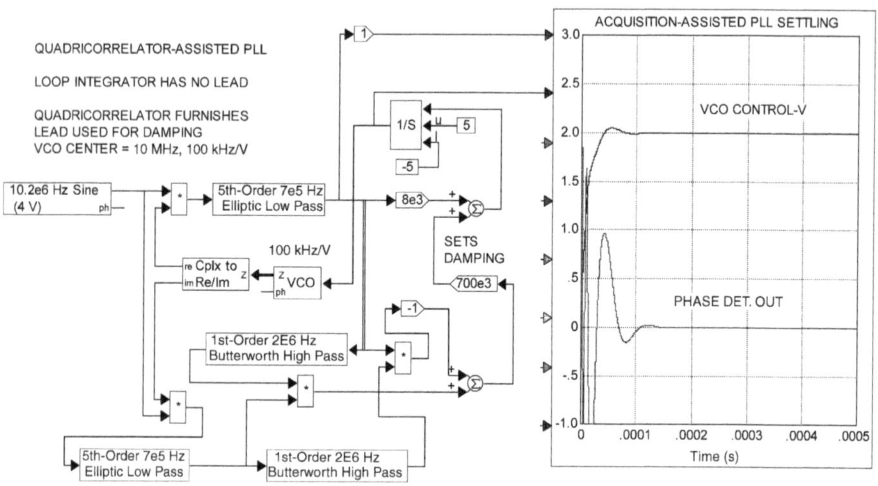

FIGURE 16.14 Quadricorrelator output provides damping in PLL because (no disconnect via dead zone or switch) (damping gain = 700e3).

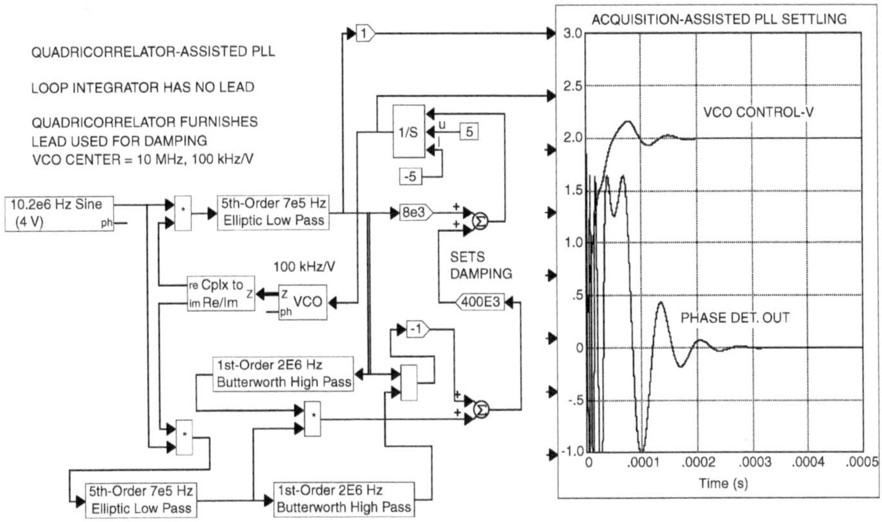

FIGURE 16.15 Quadricorrelator provides different value of damping to PLL (400e3 coefficient).

400e + 03 as in Figure 16.15 , and rerun the simulation. You will now notice a much less damped step response and much slower phase settling. The frequency acquisition loop is first order. Thus, its step response is expected to resemble an RC charging curve until the phase acquisition process occurs.

Next, load and run the file named "quadricorr_assisted_30KHzBW_PLL_sim.vsm" and the screen should look like Figure 16.16 . In this example, we have chosen

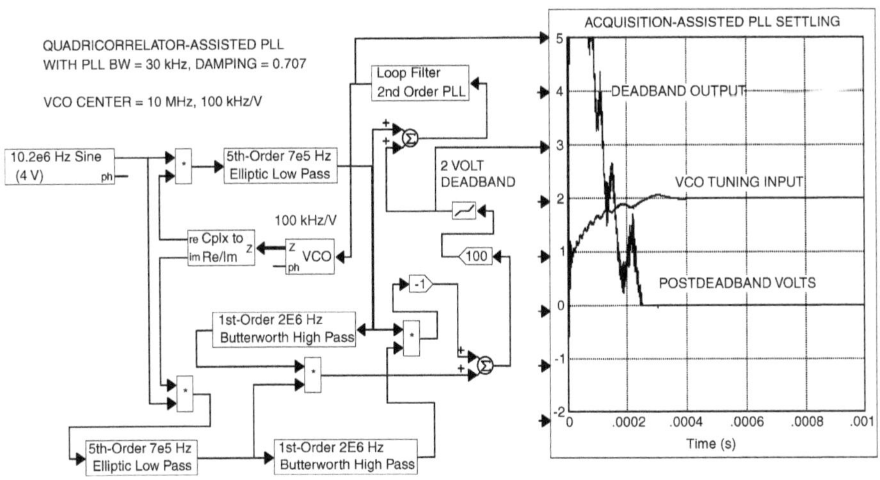

FIGURE 16.16 PLL with self-disconnecting quadricorrelator via deadband showing loop dynamics converging from frequency step.

to use a lead-lag loop filter integrator so that damping is not provided by the quadricorrelator that is self-disconnecting via a deadband. We chose a 2 V peak-to-peak deadband. Preceding this deadband is a quadricorrelator gain of only 100. This is because the VisSim model for the loop filter already has substantial gain (right click on the Loop Filter block to see its properties). The PLL BW of 30 kHz is very approximate for this example. Two waveforms are plotted: the output of the deadband and the VCO tuning control voltage.

Notice that except for the ripples, the frequency convergence (time response to a frequency step) follows an RC charge curve.

Next, let's kick the tires again. Modify the deadband and rerun the simulation. Then restore it to 2 V and modify the gain preceding it to 100e-06. This will essentially disconnect the quadricorrelator and you will see that without this assistance, the loop takes an eternity to lock (if it does so at all). Next, alter the simulation time to 5 μs and change the gain of the quadricorrelator from 200 to 50. Observe the result. A little experimentation will reward one with a great appreciation for the qualitative behavior.

A footnote may be in order to explain the modeling of a lead-lag loop integrator in VisSim. One can choose the Loop Filter block in VisSim, which does not allow for inclusion of a quadricorrelator damping contribution, or one can model the block as in Figure 16.17. The circuit at the top is the familiar type-II loop integrator with a damping resistor RF. The circuit at the bottom is fully equivalent, except for a sign inversion. Also, since VisSim does not provide for resistors and capacitors, in favor of more generality, the circuit at the bottom must be used. The great benefit is that the damping can now be apportioned between the quadricorrelator contribution and the loop integrator's lead coefficient.

So far, we have covered the basics of quadricorrelator-assisted (or any frequency discriminator assisted) PLL's. We emphasized the beauty of simulation over closed form calculations because there are usually too many trade-offs to be made between noise, damping, and the quadricorrelator's contribution to damping and how much it should be, or whether it should have no contribution once phase lock has occurred (via the use of a deadband or switch). These determinations depend entirely on the designer's unique requirements.

An additional benefit of simulation is apparent: the ability to include more high-order parasitic poles, filters, nonlinearities (VisSim provides models for deadbands, quantizers, clippers, Taylor series polynomial nonlinearities, etc.), and readouts (voltmeter, digital display, plot, spectrum analyzer, etc.), as well as the effects of noise and DC offsets, all of which would be difficult and unwieldy to include in a simplistic equation or set of equations, and even more difficult to solve while remaining sane.

If the reader is curious, he or she may wish to incorporate noise blocks within the simulation diagrams to see the effect on PLL phase (by zooming in on phase detector waveform or running an FFT on it). Then, one can perform trade-offs of how much weighting to give the frequency acquisition mechanism versus whether to remain connected or disconnected or how to apportion damping between the integrator lead and the quadricorrelator weight.

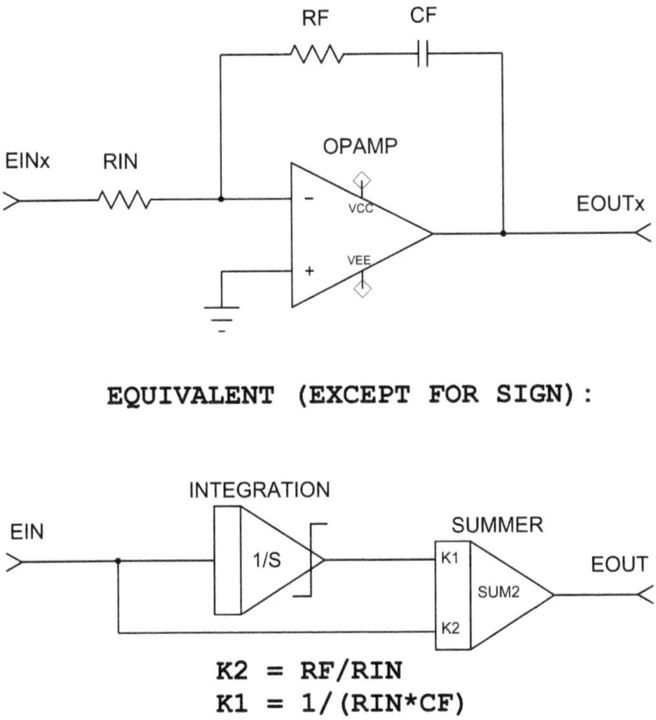

FIGURE 16.17 Equivalence of two implementations of loop integrator for a type-II second-order loop (except for a sign inversion).

16.5 COMMENTS ON SPURS' NUMERICAL INFLUENCE ON THE VCO

The VCO is a very sensitive component to external noise or spurs on both its power rail and tuning control line. It can also be sensitive to load pulling due to a noise-or-spur-modulated load impedance. That problem can be diluted by buffering the VCO. Switching power supplies should rarely be used and when they are should be postregulated with a low noise analog regulator. In some cases where phase noise is especially critical, three-terminal voltage regulators are considered barbaric for their high noise levels (100 nV per root Hz typically). How can these pollutants be assessed quantitatively for their end impact? That is the topic of this section.

Recall Equation 8.2

$$\beta = \frac{\Delta f}{f_m}$$

Also, realize that the VCO's pushing factor is P Hz/volt, and that the unwanted ripple on the VCO power rail is R volts at a frequency of f_m. Then, the DSB spectral lines at $+/-f_m$ offset from the carrier on a spectrum analyzer before the PLL loop is closed will be equal to

$$L(\text{dBc}) = -6 + 20\,log_{10}(PR/f_m) \tag{16.1}$$

for $\Delta f \leq 0.1 f_m$ and $\Delta f = PR$.

This spur level will become attenuated if it falls under the loop natural frequency f_N, by approximately 40 dB/decade. In that case, it will be equal to

$$L_S(\text{dBc}) = L(\text{dBc}) - 40\,log_{10}(f_N/f_m) \tag{16.2}$$

where the subscript S denotes suppression.

Consider an example. The power supply ripple to a VCO has 10 mV peak ripple at 20 kHz. The VCO's pushing factor is 150 kHz/volt. The PLL natural frequency is 50 kHz. From Equations 16.1 and 16.2 above, the spurs at +/−20 kHz around the carrier will be each -6-22.5-15.9 = −44.4 dBc or −38.4 dBr. This is considered good performance for many digital applications, but disastrous for radar, clock regeneration, frequency synthesis, broadcasting, and wireless applications. To improve it, we can reduce the ripple on the supply rail, superregulate the VCO rail, or increase the PLL natural frequency.

In a digital application, the spur manifests itself as jitter. The usual specification for jitter is unit intervals (UIs). One unit interval is one period of the clock (or other pulse signal).

Therefore, one UI is equivalent to 2π radians. UI can be specified in peak or peak-to-peak values. Generally, they differ by a factor of 2. Thus, jitter due to VCO pushing is equal to

$$J(\text{UI}_{\text{PEAK}}) = (PR/f_m)(f_m/f_N)^2/(2\pi) = PRf_m/(2\pi f_N^2) \tag{16.3}$$

for $f_m \ll f_N$.

In our example, jitter would have been .01∗150e3∗2e4∗6.37e-11 = 0.0019 UI. This is very good for an application like a SerDes (serializer–deserializer) where almost 0.1 UI can be tolerated.

To convert the jitter in UIs to seconds requires knowing the period of the clock signal. Suppose it is a 1 MHz clock, which is a period of 1 μs. Then, our example would have yielded 1.9 ns of peak jitter at a 20 kHz jitter rate. Had the clock (VCO frequency) been 1 GHz, the jitter would have been 1.9 ps peak. Expressing this relationship formally,

$$J_{\text{PEAK}}(\text{s}) = J(\text{UI}_{\text{PEAK}})/f_{\text{VCO}} \tag{16.4}$$

For spurs entering the VCO control line, the equations are the same but with P replaced by K_V in Hz/volt. The quantitative difference is that K_V is usually much greater than P. Typical values are 1–20 MHz/V for a 100 MHz VCO. For a 10 MHz/V tuning slope and the same jitter result as our example above, the VCO tuning line pollution must be under 150 μV. For a 100 MHz/V slope we must have pollution

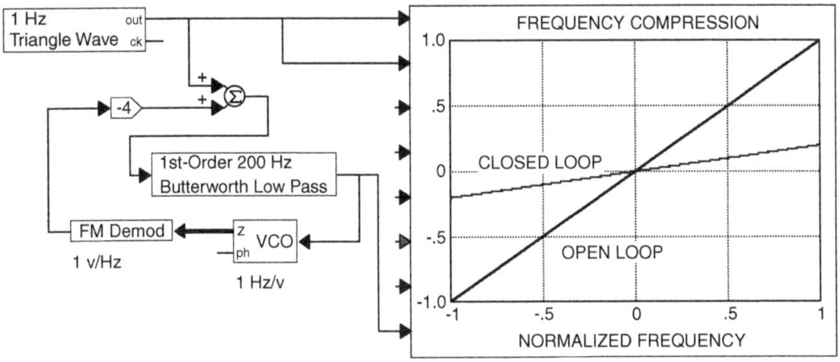

FIGURE 16.18 Example of finite frequency compression.

under 15 μV, and to improve spurs by another 30 dB we must have tuning line pollution under 0.48 μV. This illustrates why so much attention is usually paid to impedance levels (high impedance nodes are more susceptible to pickup of stray noise and spurs) and shielding.

The above equations must be modified for $f_m \geq f_N$ by setting them equal $(f_m = f_N)$. Clearly, this is not literally true, but forces the end calculation to be more accurate.

It should be obvious that low spur frequencies cause more PM than high spur frequencies due to Equation 8.2. However, the loop suppression factor improves as the square of f_m/f_N due to the closed loop PLL's 12 dB/octave high-pass nature.

16.6 FREQUENCY COMPRESSION

Occasionally, it is desired to limit the loop gain of the frequency acquisition for noise or other reasons. The result is called frequency compression, so named because a delta frequency excursion is made to shrink to a smaller excursion but not to zero error. Observe Figure 16.18 . A frequency discriminator (or quadricorrelator) of the same volts per Hertz as the Hertz per volt of a VCO is fed back via a gain (in this case an inverting gain of 4).

This shrinks the frequency excursion to $(4+1)^{-1} = 20\%$ (1/5) of the open-loop excursion, or in other words, instead of tracking out the incoming frequency error it leaves about a 20% residue. Obviously, this technique cannot be used when the quadricorrelator feeds the loop filter of a type-II PLL, since that loop filter has infinite gain at DC, not the finite gain shown in Figure 16.18. The 200 Hz low-pass pole is used to stabilize the control loop in the example. The loop must be a local loop around the VCO that does not encompass the PLL loop integrator. Gains and pole frequencies may be distributed and apportioned differently from the example given.

17

CLOCK RECOVERY USING A PLL

17.1 PLL ONLY

An important application of a PLL is the recovery of a clock waveform from a data stream. If the PLL has a finite specified high inertia, it can function as a jitter attenuating PLL or "Golden PLL." Before the PLL can be employed, however, the data must be processed to yield only the clock and sidebands. A truly random data pattern has no energy at the clock frequency because it has zero mean (it is a "balanced code" or one having a roughly equal number of high and low states). A crude method for extracting a spectral component at the clock frequency works by differentiating data waveform edges and full-wave rectifying (absolute value operation) the result. A portion of a randomized NRZ data pattern is shown in Figure 17.1. The original clock signal is shown as a series of timing ticks (impulses). Now the question is: given only the waveform in the upper trace how can we extract the clock as a continuous uninterrupted periodic signal?

One way to perform a step in that direction is to differentiate the edges of the data waveform via a high-pass filter, as shown in Figure 17.2. Then, we can full-wave rectify to get energy at the clock frequency. Prior to full-wave rectification, the recovered spectrum looks like Figure 17.3; notice that the spectrum has a null at the bit rate (clock frequency). After full-wave rectification, the recovered spectrum looks like Figure 17.4. Notice that we now have strong spectral energy at the clock frequency.

Unfortunately, the spectrum of Figure 17.4 does not yield the clock frequency alone. Data sidebands exist as well. The troubling frequencies are the ones close to

Frequency Acquisition Techniques for Phase Locked Loops, Daniel B. Talbot.

Published 2012 by John Wiley & Sons, Inc.

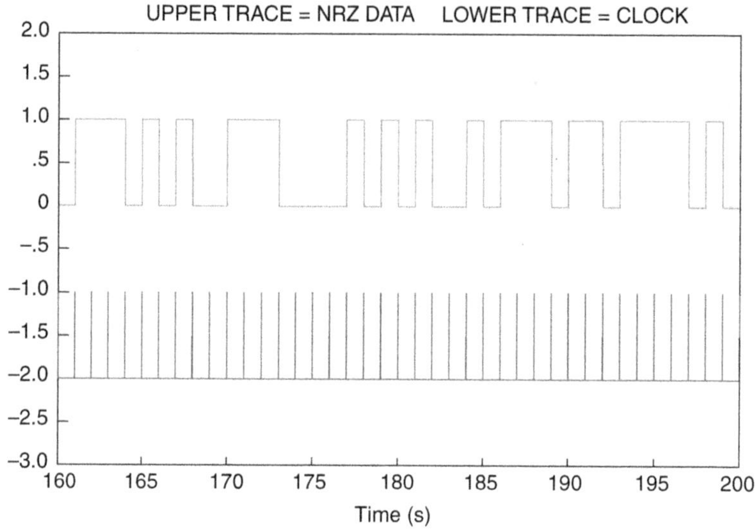

FIGURE 17.1 NRZ data and its clock.

carrier (clock), but these sidebands can be greatly reduced in severity by limiting the low-frequency code energy. If that is done, one can proceed with a narrow-band PLL that will remove the data sidebands farther away from the carrier. But before we launch into the PLL issues, let us examine a better way of extracting the clock prior to the PLL.

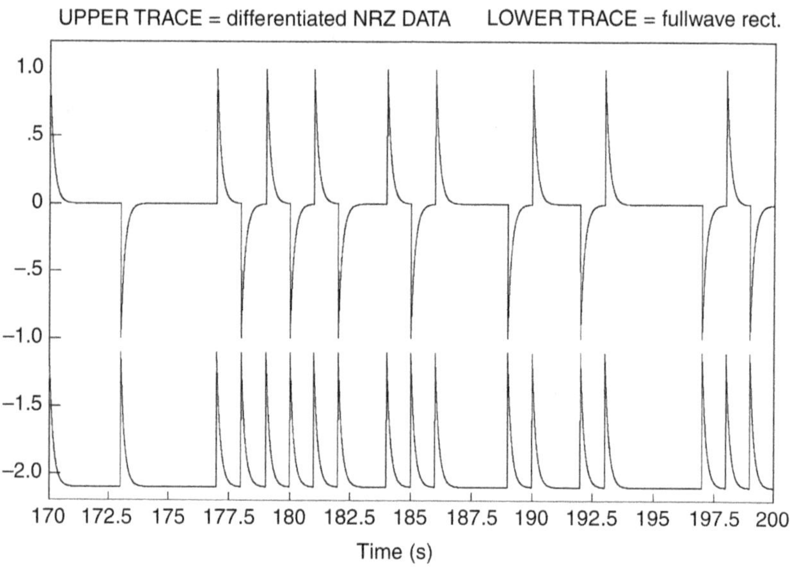

FIGURE 17.2 Differentiated NRZ waveform before and after full-wave rectification.

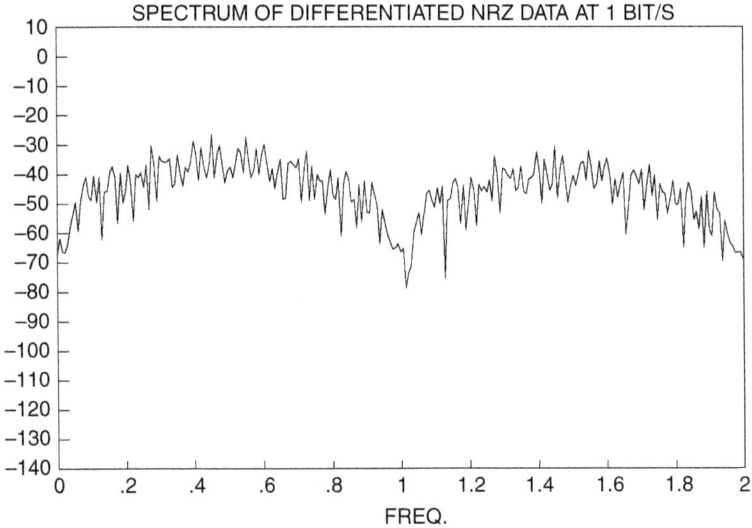

FIGURE 17.3 Spectrum of differentiated NRZ waveform.

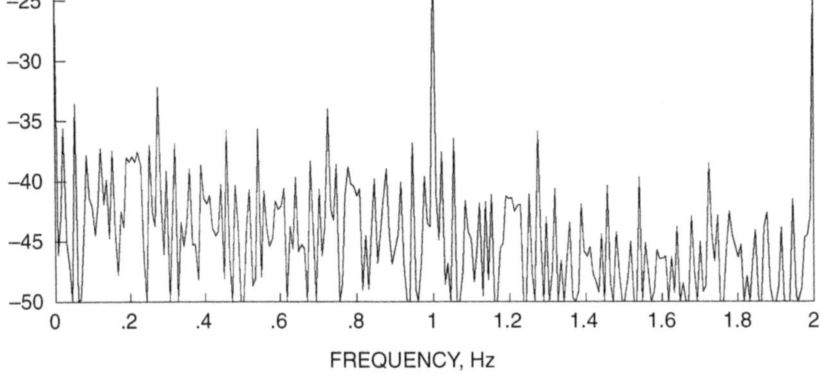

FIGURE 17.4 Spectrum of differentiated and full-wave rectified NRZ waveform.

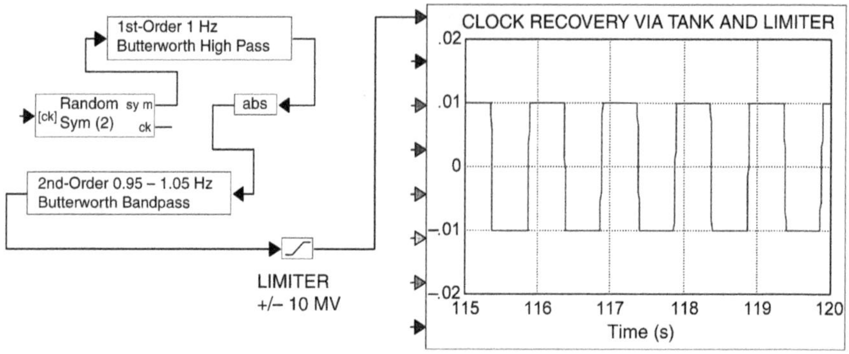

FIGURE 17.5 Time domain waveform of edge stripper followed by bandpass filter and hard limiter.

If we bandpass filter the stripped edges (after full-wave rectification), we will have injected inertia into the signal, so that some filter ringing (even though quickly damped) persists during missed edges. However, AM will result, causing needless sideband energy. But if we follow this signal with hard limiting, we remove the AM and sidebands become attenuated. If we use sufficient depth of limiting, the AM due to missed edges will be minimal. However, phase drift will occur during missed edges because the bandpass filter ringing will tend toward its own resonant frequency, which may be slightly offset from the actual clock frequency. Thus, tuning the filter is an issue. Furthermore, if the clock frequency is agile, a fixed-tuned bandpass filter will not work.

The use of the bandpass filter after full-wave rectification of differentiated data edges is shown in Figure 17.5 and the resulting recovered clock is also shown. The spectrum of the waveform is shown in Figure 17.6. Notice that the data sidebands have been greatly attenuated, leaving a purer copy of the clock signal for further purification via a PLL if desired. In some applications, no PLL is required because the spectral purity is considered adequate after limiting. The reader is encouraged to

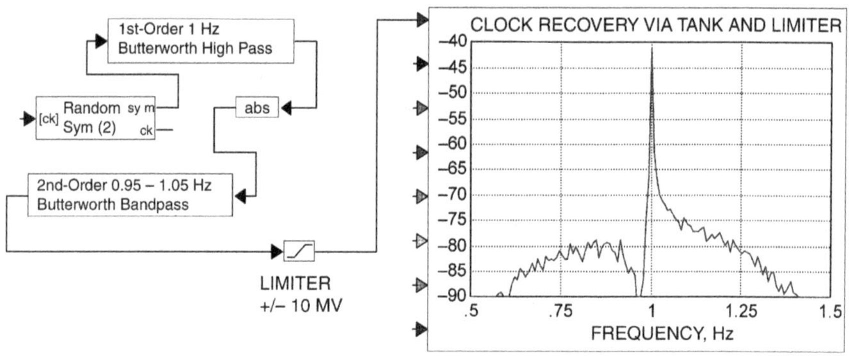

FIGURE 17.6 Spectrum of edge stripper followed by bandpass filter and hard limiter.

"kick the tires" again, by running the file named "CLOCK_RECOV_VIA_TANK_AND_LIMITER.vsm" in the VisSim subfolder in the SUPPLEMENTAL SOFTWARE FOR BOOK folder in the toolkit. Try adjusting the limiter depth (upper and lower clipping points) and observe the frequency spectrum purity close to the clock spectral line. Also, try widening or narrowing the bandpass filter.

A number of techniques exist for minimizing missed edges in the data stream: 4B/5B, Manchester, and so on. The fewer the missed edges are, the wider the bandpass filter or PLL can be, or the need for high depth of limiting decreases. If an NRZ code must be used, it should be evaluated for the potential of limiting the duration of successive ones or zeroes. This can sometimes be done with special coding techniques.

There is a much better way to strip edges from the data stream that does not require full wave rectification. A delay line and mixer (or an exclusive OR) is superior to the differentiator and rectifier method. Suppose the edges exhibit poor rise and fall times, as in the LTspice circuit of Figure 17.7. This is often caused by inadequate bandwidth or lack of (or poor) equalization. In this example, the rise and fall times are slow at 1 ns each, with a bit time of 5 ns. An alternating pattern of ones and zeroes is represented by the signal labeled PULSE at 100 MHz with voltage swings of −1 to +1 V. C1 and R1 form the pseudo-differentiator with a 1 ns time constant. This number was chosen because at the end of a bit time of 5 ns the differentiator will have discharged 82%. It is important for this memory of each pulse edge to evaporate before the next pulse occurs; otherwise, we get a problem akin to ISI (intersymbol interference). Run the file named "EDGE_STRIPPERS.asc" in the "SPICE files from author" subfolder. Click on the simulate icon to start, and the waveform plots should appear. Right click on a circuit node to see its waveform, or simply tweak the

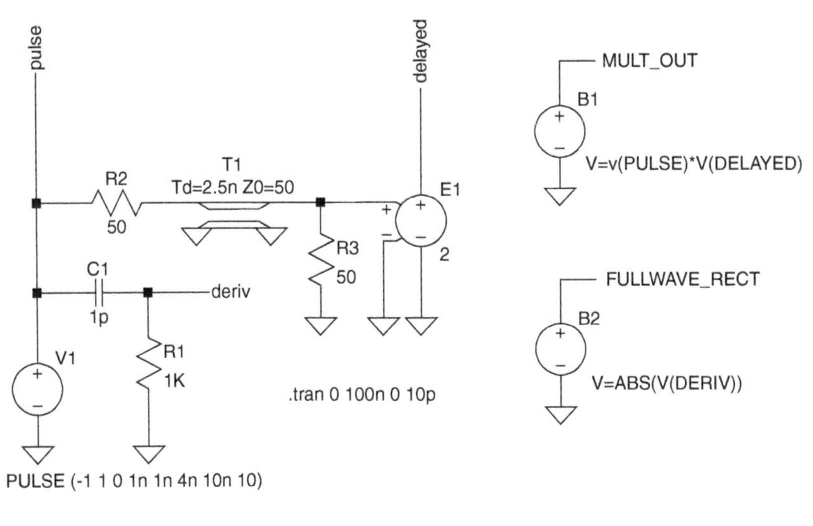

FIGURE 17.7 LTspice simulation models of differentiator-plus-rectifier edge stripper versus delay-and-multiply method.

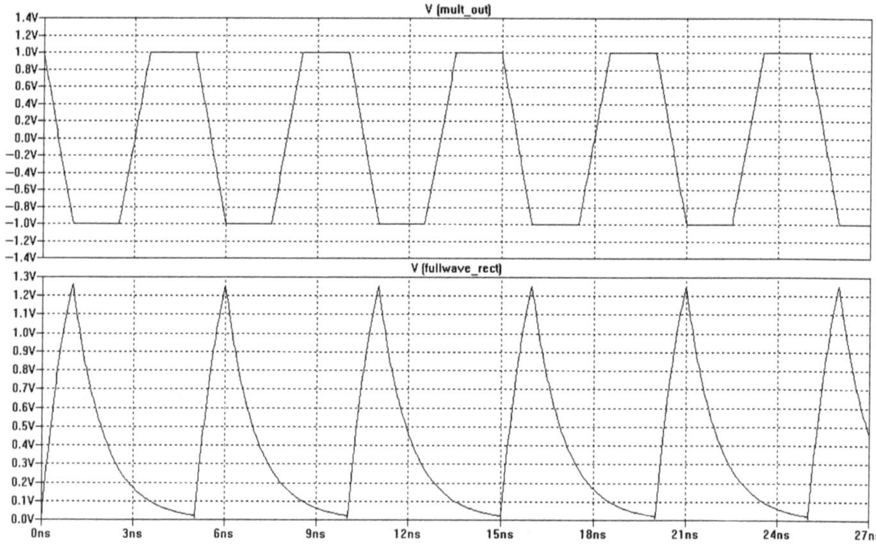

FIGURE 17.8 Waveforms comparing the efficacy of the delay-and-multiply versus differentiate-and-rectify edge stripping methods.

component values and observe the response. Notice that the delay-line-and-multiplier circuit topology outperforms the differentiator-plus-rectifier. The result is a clear amplitude advantage for the delay line circuitry. The delay is equal to one half of a bit time (1/2 of 5 ns). The waveforms are shown in Figure 17.8. The top trace is clearly stronger in fundamental clock frequency waveform purity and amplitude (being almost twice the amplitude of the lower trace and with more symmetry). In addition, the delay line method has better ISI immunity if the delay line reflection is minimal (properly terminated). The model for the delay line in LTspice requires 50 ohm terminators at each end, which throws away 6 dB of amplitude, so E1 is added to get back the 6 dB. Notice (in Fig. 17.7) the use of behavioral voltage sources B1 and B2, which allow modeling of various circuit blocks of a behavioral nature. B2 models a full-wave rectifier whereas B1 models an analog multiplier.

A "Golden PLL" for the 8.5 Gbit/s fiber channel application is shown in Figure 17.9. It is designed to recover the clock and attenuate jitter according to a specific jitter tolerance and roll-off. It must add extremely low jitter of its own, on the order of < 0.01 UI. It outputs a clean 8.5 GHz sine wave clock that has been processed by a PLL and a cavity filter. It employs the delay line and multiplier method of edge stripping in order to maximize SNR prior to PLL processing and to yield robust SNR in the presence of significantly shallow edge slopes due to nonflat frequency response of the applied signal's bit stream. It also features another 8.5 GHz cavity filter immediately following the edge stripper, in order to ignore missing edges as much as possible. This cavity filter is then followed by a low noise hard limiter of > 10 dB limiting depth. The cavity filters are each 100 MHz wide (three cavities each). The subsequent PLL is a sweeping acquisition type swept by a

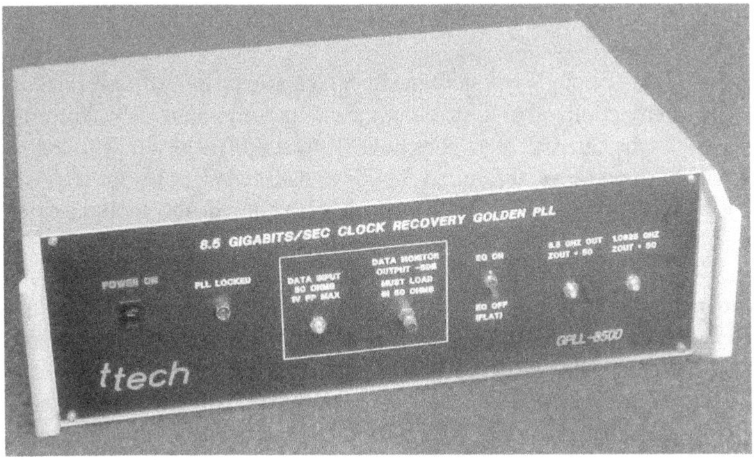

FIGURE 17.9 A "Golden PLL" laboratory clock recovery PLL instrument for 8.5 Gbit/s fiber channel precision applications.

built-in 3 Hz sine wave generator with coherent lock detector and switched disconnect. The edge stripper is shown in Figure 17.10, and it employs a wideband mixer and adjustable phase trimmers (the large tubular objects) to obtain a delay difference of $^1/_2$ the period of 8.5 GHz or 58.82 ps (approximately 0.7 in. in air, less in cable). The effective delay of the semirigid cables and the internal delay difference of the LO and RF ports of the mixer (multiplier) are thus tuned out by the adjustments and the $^1/_2$ bit delay is established accurately. The alignment procedure consists of using a modulated swept signal and an oscilloscope to find the $^1/_2$ bit point.

FIGURE 17.10 A tunable delay line-plus-mixer edge stripper for 8.5 Gbit/s fiber channel PRE-PLL signal conditioning.

17.2 PLL WITH SIDEBAND CRYSTAL FILTER(S)

A crystal filter may be of great benefit in either cleaning up phase noise sidebands in a synthesizer reference signal or in sustaining the edge stripper clock waveform during missing edges. One can realize a "poor man's" crystal filter by simply using a cheap series resonant quartz crystal (around $2.00) to pass the signal from its source to a load. See Figure 17.11 and the included LTspice file (in the toolkit folder titled SUPPLEMENTAL SOFTWARE FOR BOOK\Spice files from author) named "10MHZ_XTAL_FILTER.asc." This example uses a 10 MHz series-resonant *fundamental* mode crystal represented by lumped elements L3, R4, and C3 and parasitic shunt capacitance C4 (on left side of diagram) to yield a noncompensated result "OUT_NONCOMP" shown in the lower frequency response plot of Figure 17.12. The parasitic can be optionally negated via an inductor external to the crystal, L2 in the diagram. The internal model for the crystal is the same as before, now represented by L1, R2, and C1, and parasitic shunt capacitor C2. The new output is "OUT_COMP" and the response is shown in the upper portion of Figure 17.12. The compensation inductor L2 may have significant shunt capacitance of its own, which will force a lower inductance value to be used. The circuit values used here result in 6.85 dB of loss, which can be reduced by using a crystal having lower internal series resistance. The bandwidth is affected by all resistances, and a load resistor of more than 150 ohms is probably not optimal. Again, the reader is encouraged to "kick the tires" and vary the resistances and ratio of C3 to L3 (affecting Q). A cheap crystal may yield only a moderate Q (around 2000–10,000) and may also have acoustic spurs (multiple parasitic resonances) and most crystals will exhibit responses at overtone frequencies. A quartz crystal is after all a piezoelectric (acoustic and electric) device. Before designing it into a product, it should be fixtured as in Figure 17.11 and swept for its frequency response and spurs.

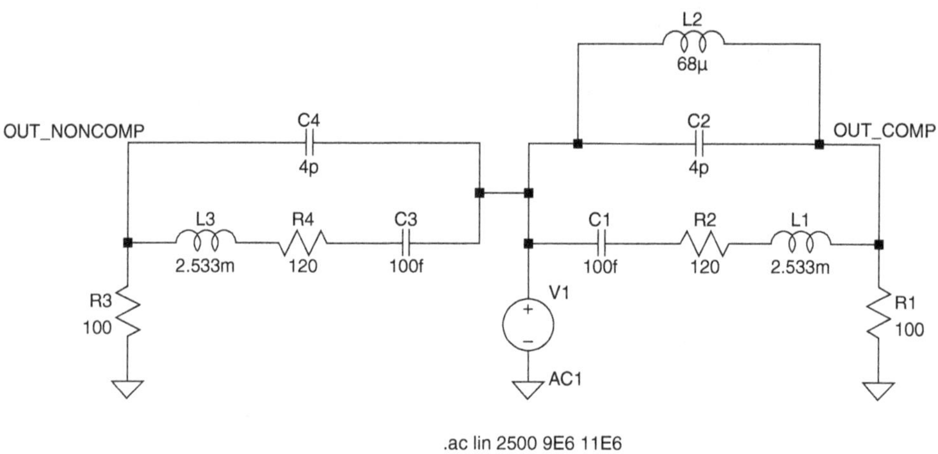

FIGURE 17.11 "Poor man's crystal filter" for 10 MHz (LTspice models).

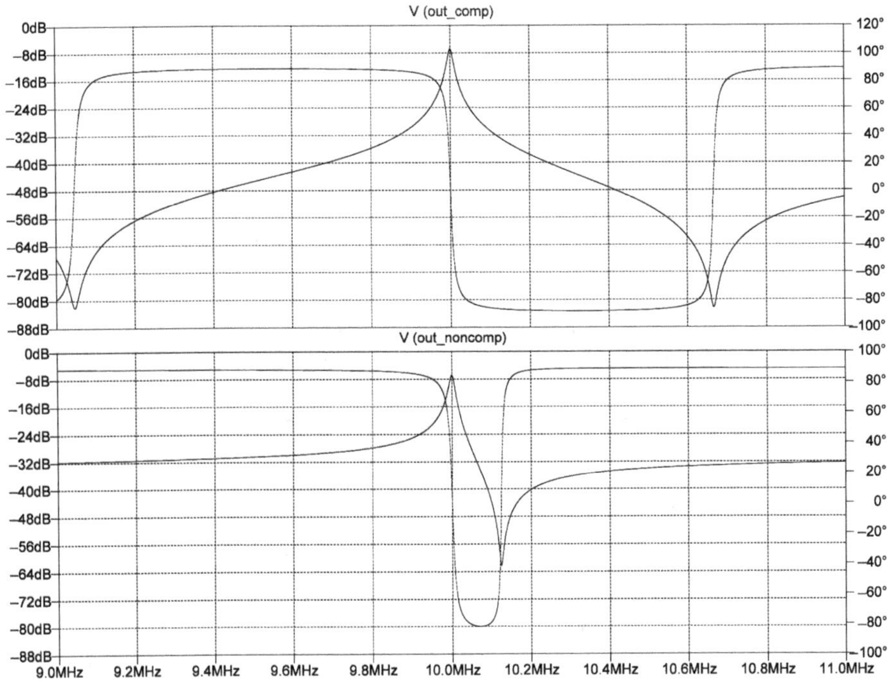

FIGURE 17.12 Frequency response for crystal filter example of Figure 17.11.

The SPICE model does not include spurious mode modeling. Incidentally, LTspice can replace the detailed separate RLC crystal model with a single capacitor model. Try right-clicking on the capacitor and you will see that it can model a crystal all by itself (there is already provision for the series inductance and resistance as well as the shunt capacitance). So, we need not draw all the individual components of the crystal. Since the crystal is used series-resonant, increased Q is obtained by increasing the ratio of L3 to C3 (smaller C, larger L). This is opposite to a parallel resonant circuit, where higher Q is obtained by having larger C and lower L.

The crystal filter is generally applicable to PLL's operating under 100 MHz or so. It is useful for its highest Q for lower frequencies. The actual PLL design is similar in approach to the PLL design for high frequencies using a cavity filter that is described next.

17.3 PLL WITH SIDEBAND CAVITY FILTER

The 8.5 Gbits/s "Golden PLL" previously described uses cavity filters each as shown in Figure 17.13. The frequency response (and return loss) is shown in Figure 17.18. A cavity filter has extremely low insertion loss (IL = 2 dB in this case) combined with

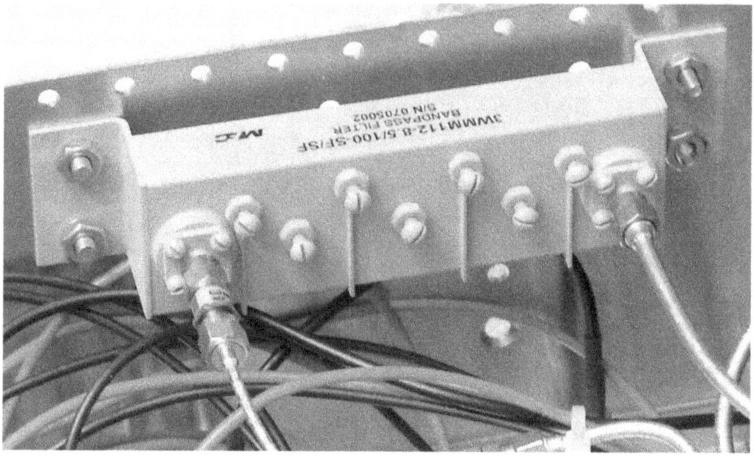

FIGURE 17.13 Picture of 8.5 GHz cavity bandpass filter, three-section, 100 MHz BW.

very low percentage bandwidth (BW divided by center frequency as a percentage), owing to its high Q resonator sections. It is essentially a coupled resonator filter that we have mentioned in previous discussions. Its temperature stability of center frequency depends on material selection and machined parts' temperature characteristics.

At high frequencies it is often necessary to employ isolators to achieve low noise impedance matching and high reverse isolation. An isolator is a nonreciprocal device that transmits signal primarily in one direction (from input to output) and substantially blocks signal transmission in the reverse direction (output to input). The reverse isolation is typically 20 dB, and the forward insertion loss is usually around 1 dB or less. Power handling is much greater than a crystal filter, but an isolator becomes physically large below 200 MHz. Often, an isolator is built using a circulator with port 3 terminated. At very low frequencies, lumped reactance-based designs called isoductors exist. These have gone out of fashion since they are in the same frequency range as unilateral active buffer amplifiers. The buffer approach also has better isolation, although at a penalty of higher noise and lower power handling, and a requirement that DC power be supplied. To deliver 1 W RMS into 50 ohms, the minimum bias current for a buffer must be 200 mA, as compared to 0 mA for an isolator of similar RF power handling. (The peak current into the load cannot be greater than the bias current; therefore, to deliver 1 W into 50 ohms requires a buffer capable of 10 V peak, which is 200 mA. In reality, it must provide a 50 ohm output impedance also to match the load, so it may need to provide 20 V peak into 100 ohms, which is still 200 mA.)

A wideband isolator for 8.5 GHz is shown in Figure 17.14.

The block diagram of the 8.5 GHz golden PLL is shown in Figure 17.15. Notice the presence of several stages of isolation, including a buffer stage internal to the VCO. It is necessary that the edge stripper output does not injection lock the VCO, especially since the two signals are at the same frequency. The reference (edge

FIGURE 17.14 Picture of an 8.5 GHz RF isolator (port 3 terminated in 50 ohms).

stripper) signal must not even injection pull or influence in any significant way the phase of the VCO. Otherwise, the VCO might tend to follow the jitter (or lack thereof) of the reference signal.

Because of the high input return loss (low reflection coefficient) requirement, and the requirement for pass-through for monitoring, a 6 dB loss resistive splitter was used to connect the input data stream signal to the subsequent device (a 3 dB attenuator to improve return loss still further and isolate the subsequent stage). A bypass switch exists to select either "FLAT" or "EQUALIZED" input signal frequency response. The bypass or FLAT mode is self-explanatory, but the EQUALIZED (EQ ON) mode sends the signal through a 2500 MHz high-pass filter. Its function is to drop the elevated spectral components when a downward sloping spectrum is applied. The name "EQ" is not strictly correct, but this mode can help to dig the signal spectrum of interest out of competing strong low-frequency energy. Subsequently, the signal is applied to a high gain amplifier and 3 dB in-phase power splitter, the output of which feeds the delay line-and-mixer edge stripper. The edge stripper output drives a cavity filter for reasons previously discussed, and the output of that cavity filter drives a special microwave (2–12 GHz) amplifier having very low even-order nonlinearity. It is also a good limiter when overdriven. In this product, it is overdriven by 10 dB or more (10 dB depth of limiting) for a normal input signal level (500 mV peak to peak). Its output is quite strong when in limiting, so there is subsequently an attenuator stage to prevent burnout of the next stage(s). Following the attenuator is a 90° power splitter and its outputs drive an orthogonal (I and Q) mixer. The 90° Q output drives the PLL loop integrator while the in-phase I output serves to recognize when phase lock has occurred in order to disconnect the sweep signal (a 3 Hz sine wave source that always runs; such a low frequency was chosen because the disconnect switch does not possess infinite ON/OFF ratio, so any sweep signal leaking into the loop integrator should be as low in frequency as

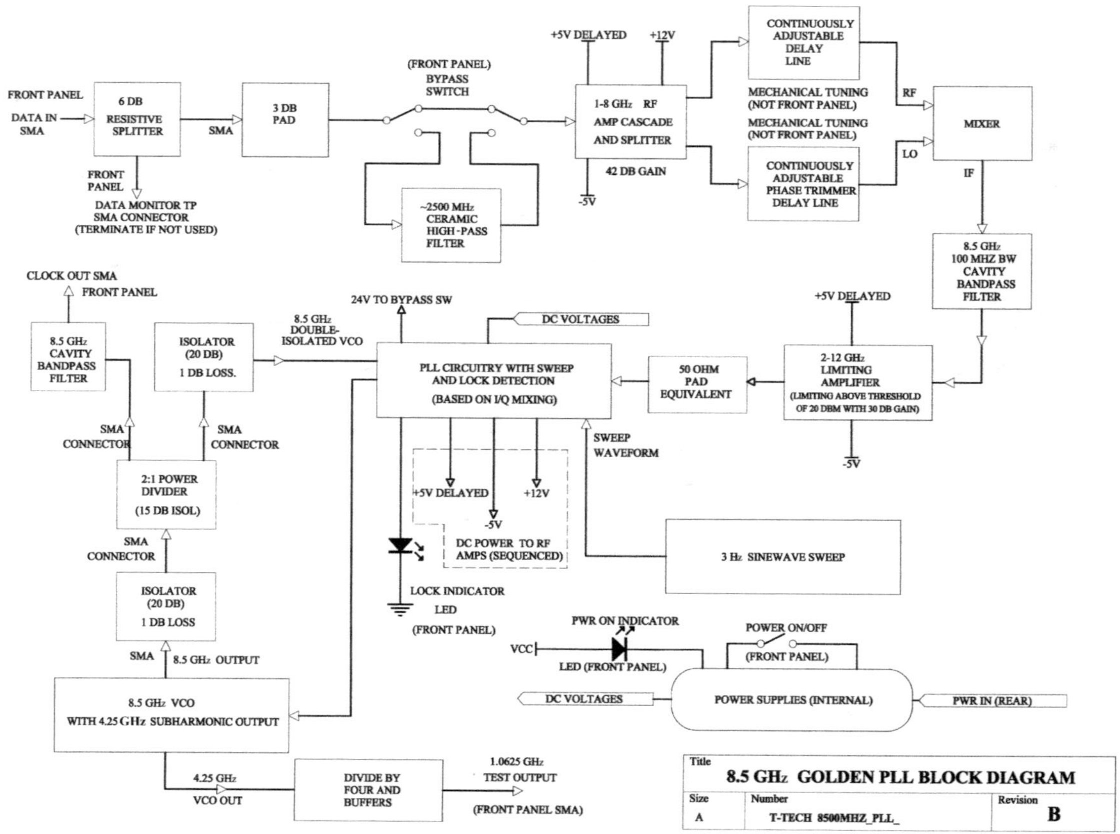

FIGURE 17.15 Block diagram of 8.5 Gbits/s golden PLL for fiber channel applications (e.g., jitter measurements).

practical (the PLL has huge attenuation of a low-frequency disturbance as discussed in previous chapters). Care is exercised in attenuating harmonics of the sweep signal for the same reason.

To prevent false locking on data sidebands, the swept acquisition and tracking range of the VCO is limited to less than +/−50 MHz by a simple trick to be described soon.

The center frequency of the VCO sweep range is set via a trimpot, and that circuitry will be discussed shortly.

The VCO output feeds a cavity filter to remove far away noise sidebands, pattern jitter, and harmonics. The output of the cavity filter then appears at the instrument's front panel. A front panel test point is also provided that represents the 8.5 GHz output signal divided by 8. This was desired for phase jitter testing due to instrumentation availability. It is not intended to be used as the recovered clock, although it will have one-eighth the pattern jitter of the 8.5 GHz output.

Next, let's look at some of the key circuitry in detail. Refer to Figure 17.16 (also viewable in the toolkit as a TIF file). A cheap (less than $1.00) quad of comparators (U4 A, B, C, and D) forms a window comparator operating on the coherent lock detector I mixer output after filtering (not shown). When the I mixer output is greater than a certain minimum absolute value, the quad comparator circuitry opens the sweep switch (U5) disconnecting the sweep signal from the loop integrator op-amp (U7) and illuminating the lock indication LED (light emitting diode, D13). The Q mixer output (after filtering, not shown) feeds the loop integrator via trimpot R62 and series resistor R65 (which limits the adjustment range). The setting of trimpot R62 establishes the PLL BW. The PLL is highly damped so as to appear like a low-pass response with no peaking. The damping is set via R54.

The sweep and acquisition span limiting spoken of earlier is simply accomplished R21 in cooperation with R70 and R71. These resistors attenuate the DC swing from op-amp U7, while AC and DC feedback is taken from the load side of R21 (thus, it is enclosed in the loop integrator feedback path). Capacitor C36 assures that AC gain of op-amp U7 is not spoiled at very high frequencies.

The sweep limiting attenuator just described contains a load (R70 + R71) whose return node is a variable baseline established via trimpot R66, thus making the sweep center frequency adjustable. The center tap point of R70 and R71 feeds the VCO, and the limits are approximately +/−150 mV around the baseline established via R66. Capacitor C34 (in cooperation with R28 and the tapped equivalent resistance of R66) filters out the AC noise from regulator U14 so that the VCO baseline is essentially noiseless. There is provision to superimpose an AC test signal from J2 upon the VCO baseline. This enables the calibration of PLL BW by observing the response at J1 to the test signal sweep at J2. Since R70 = R71, the sweep test signal and its response at J1 will be equal (and opposite in phase) at low frequencies (below the PLL BW). J1 is terminated by R29 to feed a 50 ohm coaxial cable, but must not be loaded to ground at DC or AC by a resistance less than 10 k ohms. This requirement allows for connection to a high-impedance AC decibel meter or oscilloscope.

To insure insignificant VCO pushing, the VCO power rail is superregulated by circuitry shown above J1 (components beginning with R25 on the left and ending with J3 on the right). Zener D9 insures that upon power supply sequencing

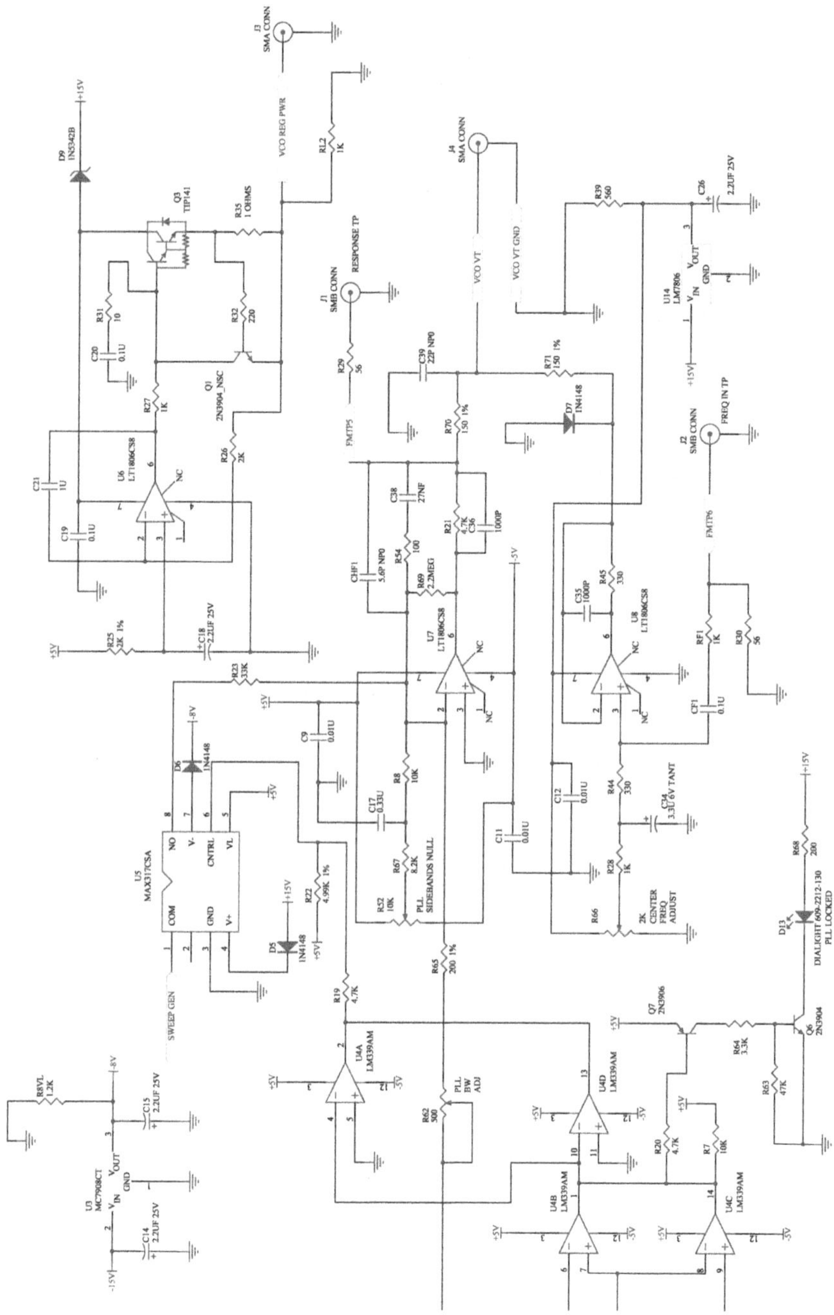

FIGURE 17.16 Key PLL circuitry schematic diagram for golden PLL.

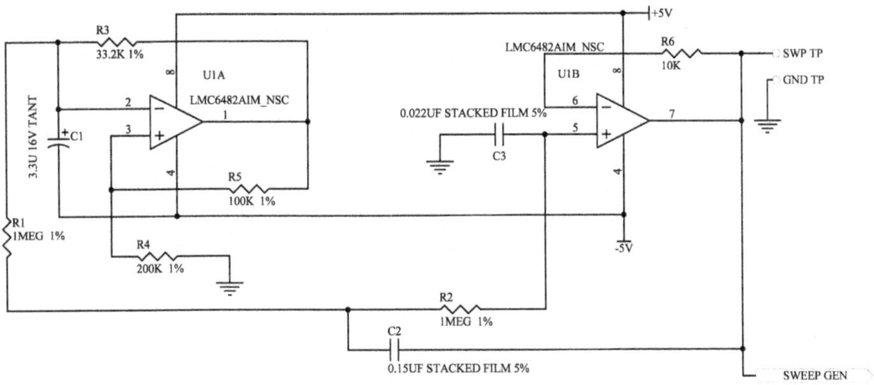

FIGURE 17.17 Schematic diagram of a 3 Hz sine wave sweep generator.

(at turn-on), the DC voltage spike (if any) at J3 will not burn out the VCO. Resistors R35 and R32 operating with Q1 limit the VCO load current at J3 to about 600 mV, protecting the superregulator from brief short-circuiting at J3. Resistor RL2 assures that enough current (5 mA) will load the superregulator circuit to bring transistor Q3 into a reasonable portion of its beta-curve when added to the VCO load current, and to allow reasonably significant loading for proper operation for testing the superregulator prior to attaching the VCO power cable. This is advisable in case the superregulator is outputting excessive voltage that might damage or stress the VCO, due to a fault before final assembly.

The sweep source is a 3 Hz sine wave generator shown in Figure 17.17. The LTspice simulation model is provided in the toolkit as "3HZ_SWEEP_GEN.asc" and the waveform is graphed in the accompanying plot.

The cavity filter of Figure 17.18 helps to remove data pattern sidebands on the output clock from the "Golden PLL" in addition to low-pass filters after the I and Q phase detectors, but obviously the attenuation is not infinite. When data edges are missing, we would like the phase detector to output nonoffset zero-signal baseline. If there is any DC offset, it will appear as a pulse when edges are missing, thus causing unnecessary sideband energy. This offset can be nulled via trimpot R52 in Figure 17.16, and the adjustment can be made while observing sidebands on a spectrum analyzer and adjusting for a null. Or, the adjustment can be made by observing sidebands while feeding a pulsed sine wave of 4.25 GHz into the Golden PLL data input port. To permit better visibility of these sidebands, the pulse repetition frequency (PRF) can be low (around 2 MHz) with a duty factor of about 5% in order to sustain the PLL BW (phase detector effective gain is dependent on duty factor of the pulsed RF sine wave). The ON-to-OFF ratio of the pulser should be >30 dB.

The combination of all these features and techniques yields a true "laboratory-standard" instrument. We demonstrated them to give an extreme example of PLL clock recovery quality. Often, as in a clock recovery IC design, the goals are much lower.

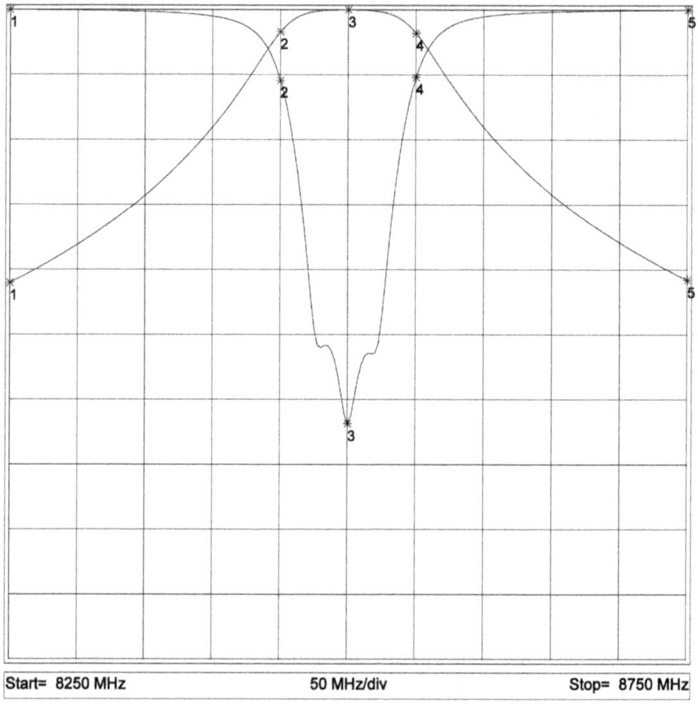

FIGURE 17.18 8.5 GHz cavity filter frequency response (top trace at 10 dB/div scale) and return loss (bottom trace at 5 dB/div scale); not shown is 2 dB I.L.

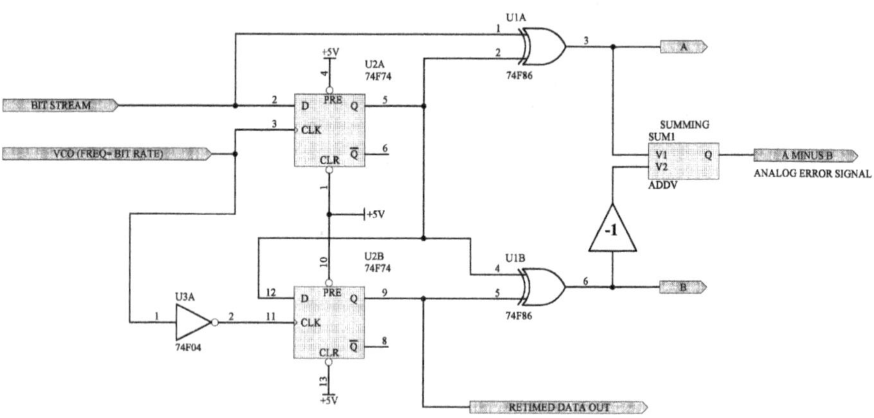

FIGURE 17.19 The Hogge phase detector (example using 74 F series logic).

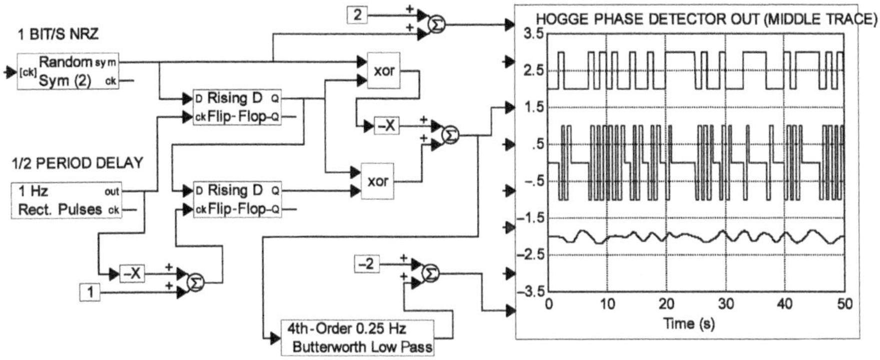

FIGURE 17.20 Hogge phase detector showing NRZ data (top trace), phase detector output (middle trace), and output average (via filter, bottom trace).

17.4 THE HOGGE PHASE DETECTOR

A very interesting digital phase detector for data clock extraction is the Hogge phase detector, shown schematically in Figure 17.19, employing the "74F" logic family. Of course, the topology can be replicated using CML (current mode logic) or designed into an IC. The Hogge detector outputs zero volts (differentially for A–B) during missing edges. See Figure 17.20. It can also retime the data (a feature that provides a jitter-attenuated data bit stream from the original stream), as shown in Figure 17.21. Simulations are in the VisSim folder under "hogge∗.vsm." The Hogge phase detector is *not* a PFD (phase frequency detector) and requires one of the frequency acquisition techniques previously discussed. An ensemble of logic devices from the Hittite company is shown in Figure 17.22 that yields a very high-speed Hogge phase detector.

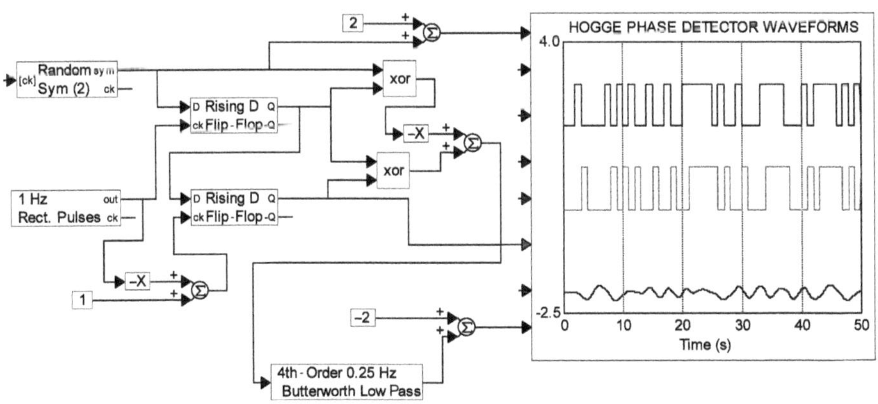

FIGURE 17.21 Example of Hogge phase detector retimed (regenerated) data output (middle trace).

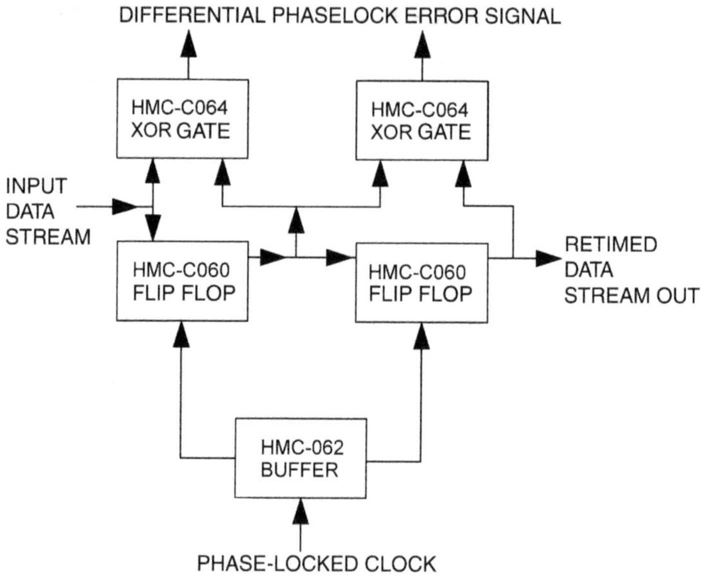

FIGURE 17.22 Example of a very high-speed Hogge phase detector from commercially available components (Hittite®).

17.5 BANG–BANG PHASE DETECTORS

Other types of phase detectors exist that have the equivalent of infinite gain followed by hard limiting. The Alexander phase detector is one example. Like most bang–bang transducers, the effect on loop dynamics and design is considerable. For this reason, we have opted to exclude them from consideration for high-performance clock recovery PLL discussion. See Figure 17.23 for the VisSim model of a bang–bang phase detector-based PLL. Run the simulation, named "BANG-BANG_PLL.vsm" in the toolkit in the VisSim folder, and tweak the "knobs" to see what happens. Notice that the PLL settles to the correct frequency for the parameters chosen in Figure 17.23. Even when damping is raised to permit capture and prevent false locking, the jitter is significant, because the bang–bang rattling of the phase detector appears superimposed on the VCO control voltage. As mentioned in a previous chapter, however, pedestrian performance is sometimes acceptable. Rarely would a bang–bang PLL be usable in a doppler radar system, for example, where phase noise is a huge issue, but it is often found in digital clock circuits where jitter is of merely minor importance.

Removing a low-pass filter following the hard-limited phase output prior to the loop integrator and reducing damping results in a cleaner view of the performance (file name "BANG-BANG_PLL_WITHOUT_LPF_ON_PLL.vsm" in the VisSim folder), as shown in Figure 17.24, but notice the phase detector output "hunting." As shown, this hunting is equal to about +/−0.9 peak degrees, which, if it were a sine

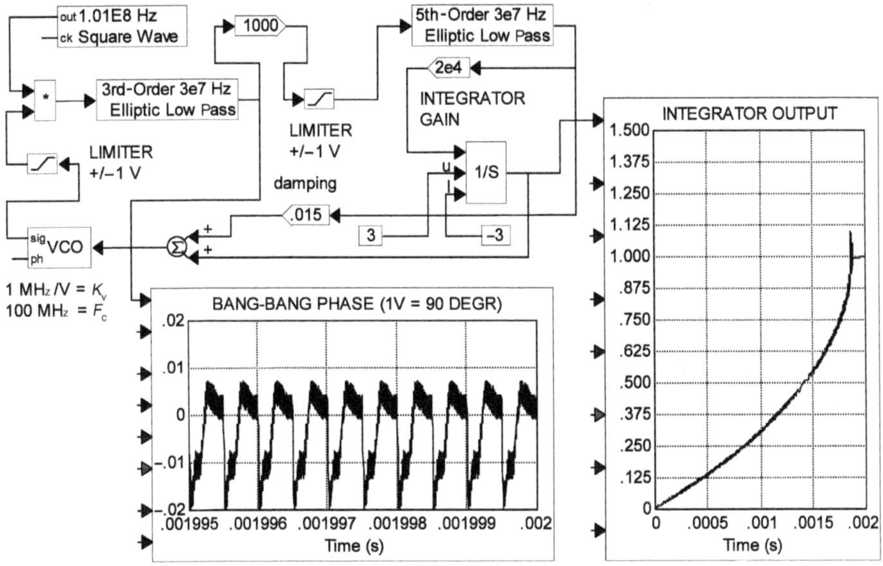

FIGURE 17.23 Example of a bang–bang PLL simulation using VisSim.

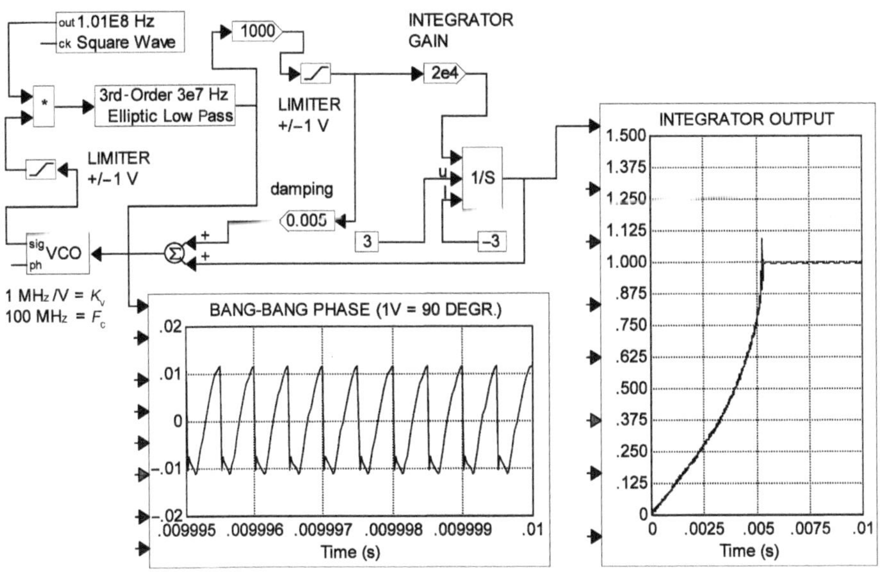

FIGURE 17.24 Bang–bang PLL omitting the second (2F) carrier filter and reducing damping, showing the bang–bang hunting actual phase angle (notice longer capture time).

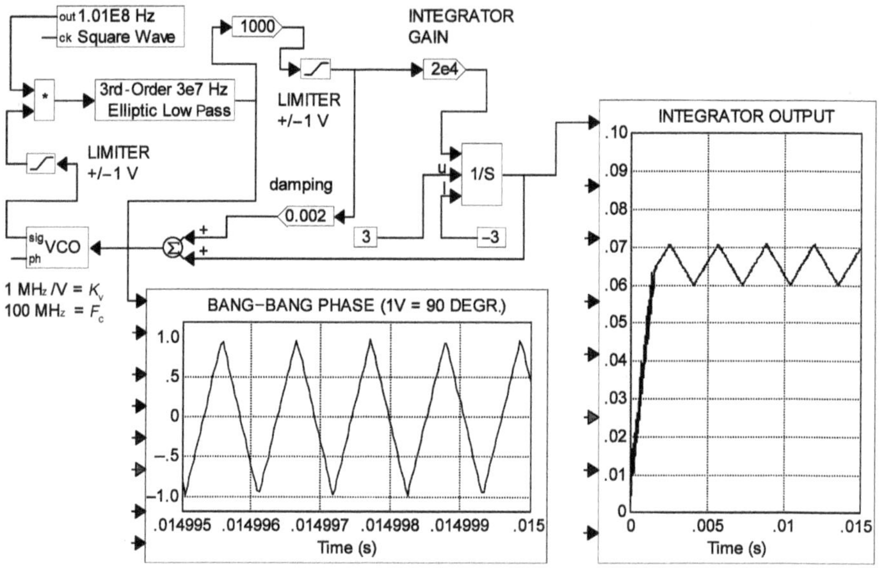

FIGURE 17.25 False lock condition from too little damping (notice out-of-lock phase beat).

wave, would yield a pair of sidebands around the carrier each at −42 dBc, or a total of −36 dBr (decibels relative to a radian) or 0.0025 UI peak. This is probably OK for a SerDes, but represents very high deterministic jitter for most other applications.

Using improper parameters (inadequate damping) results in settling to the wrong frequency (false lock), as shown in Figure 17.25, and the actual phase (not the bang–bang hard-limited phase detector output) is triangular at the beat frequency. The hard-limited phase would, of course, be a square wave at the beat frequency.

18

FREQUENCY SYNTHESIS APPLICATIONS

An important application of the PLL is frequency synthesis, which is the generation of a new frequency from a reference frequency. In the past few decades, a technique has become popular for synthesis of new frequencies without the use of a PLL, called DDS (direct digital synthesis), which has the ability to generate any frequency within its capability at very high-frequency density (closely spaced increments). Prior to that, another technique, analog direct frequency synthesis, existed, which also did not need a PLL. Direct synthesis of either type can output a very low phase noise signal from a low noise reference. DDS, however, outputs a dense spur floor, potentially having a spur every Δf where Δf represents the frequency resolution, and for a 32-bit DDS using a 1 GHz reference clock, $\Delta f = 0.23283$ Hz. Some DDS and fractional-*N*-based instruments have resolutions in the micro-Hertz range (e.g., the HP 3325B "frac-*N*" signal generator). The PTS 040 from Programmed Test Sources® can generate any frequency from 0.1 MHz to 40 MHz with $\Delta f = 0.1$ Hz spacing between possible frequencies, and uses a combination of analog direct synthesis and DDS. Hybrid synthesizers are becoming more plentiful. Some of those employ PLL and DDS techniques.

A novel approach to frequency synthesis is called RAFS (rational approximation frequency synthesis), featured in the SG380 instrument from Stanford Research Systems® in which a high phase detector frequency is used without fractional-*N* feedback and in which the reference frequency is adjusted to permit this high phase detector frequency for all possible output frequencies.

Another technique for generating a high-resolution frequency operating range is fractional-*N* (aka "frac-*N*"). In this technique, the feedback counter is not limited to

Frequency Acquisition Techniques for Phase Locked Loops, Daniel B. Talbot.

Published 2012 by John Wiley & Sons, Inc.

dividing by integers, but extends to fractional numbers by dividing by cyclic alternating numbers and then relying on loop dynamics and noise shaping (dithering) to attenuate the resulting spurs. For example, the Skyworks® SKY74038 is an IC capable of fractional-*N* synthesis up to 2.6 GHz when used with suitable accompanying components. Many vendors of ICs provide frac-*N* chips, including Analog Devices®, National Semiconductor®, Peregrine Semiconductor®, Fujitsu®, Philips®, and others.

This chapter will confine its discussions to standard PLL approaches to frequency synthesis, including *N*-over-*M* synthesis in which the reference and feedback counters are chosen to have integer values whose ratio yields the desired frequency to within an error envelope specified by the user (see the file "N_OVER_M.EXE" in the "Handy tools" folder of the toolkit). We will also examine the drift canceling or Wadley loop that permits expansion of synthesizer frequency span.

18.1 DIRECT FREQUENCY SYNTHESIS WITH WADLEY LOOP

The Wadley loop (also called "drift canceling loop") is an ingenious example of frequency compression, invented by Dr. Trevor Wadley [31]. Used in frequency synthesis, it enables expansion of the frequency span while preserving resolution and requiring only a free-running VCO (although phase locking the VCO can be an option).

The principle is illustrated in Figure 18.1. A 400–450 MHz frequency synthesizer having a fine resolution (chosen by the designer) is only capable of a 50 MHz span. With the addition of a Wadley loop the span is increased to 1.05 GHz. The technique is explained as follows. First, a comb generator is provided having a pump frequency equal to the limited frequency span that we wish to expand. (Alternatively, the comb generator can be replaced with a synthesizer of PLL or direct type.) This comb pump (50 MHz) is derived from the synthesizer's primary crystal reference (10 MHz) via a quintupler. Then, the spectrum of the desired comb harmonic lines is bracketed with a bandpass filter (passing *N* times 50 MHz where *N* is integers from 6 to 12). The comb then feeds a mixer and is mixed with the free-running VCO that is tuned in 50 MHz steps from 2.8 to 3.8 GHz. These 50 MHz steps do not need to be precise, since the Wadley loop "eats" the error for small frequency errors, as we shall see. The lower mixer in Figure 18.1 outputs a 3.4 GHz single frequency, with a small error due to imprecise VCO tuning. The 3.4 GHz filter is a "heroic filter," that is, it must output 3.4 GHz and suppress unwanted sidebands at increments of 50 MHz from the comb. This 3.4 GHz signal is then used by the middle mixer to upconvert our original 400–450 MHz synthesizer to 3.5–3.85 GHz and the result is bandpass filtered. That result is then mixed with the VCO output in the top mixer of Figure 18.1. Mixing spectral polarity is chosen so that the VCO frequency is subtracted from the 3.8–3.85 GHz signal. In this way, the VCO frequency error is canceled as a common-mode signal frequency error. The error can be represented as Δ in the figure. It is profound to realize that low-frequency phase noise (noise frequencies close to the VCO carrier) is also canceled to a degree that is quite impressive. It turns out that the delay of the

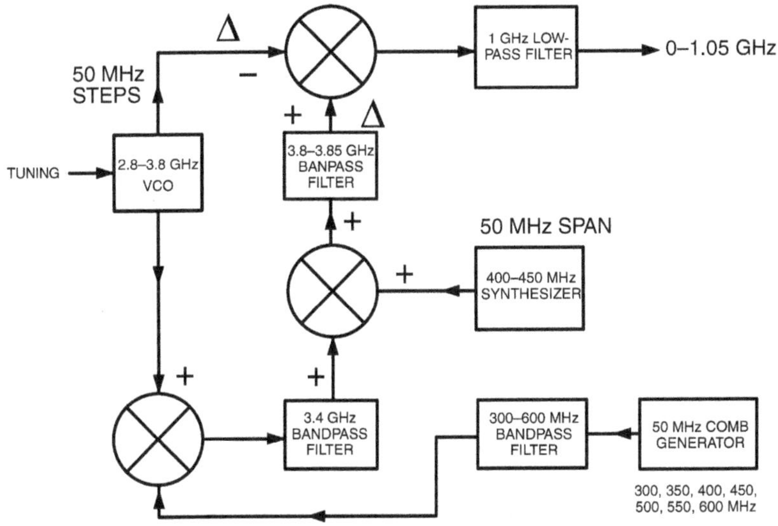

FIGURE 18.1 Block diagram example of a Wadley loop.

3.4 GHz (in this example) heroic bandpass filter dictates the limitation on the degree of cancellation. Finally, the desired output of 0–1050 MHz is obtained from the top mixer, with very little added noise (assuming the comb generator noise is the noise floor).

Figure 18.2 will show how the delay of the heroic filter affects noise cancellation in the Wadley loop. We can model the phenomenon of Wadley cancellation by a bandpass-to-low-pass transformation. We have chosen a 50 MHz 6th-order Chebyschev (representing a 12th-order bandpass or 6 resonators) to simulate the cancellation delay (in this case 16 ns) in the loop. This low-pass filter is the

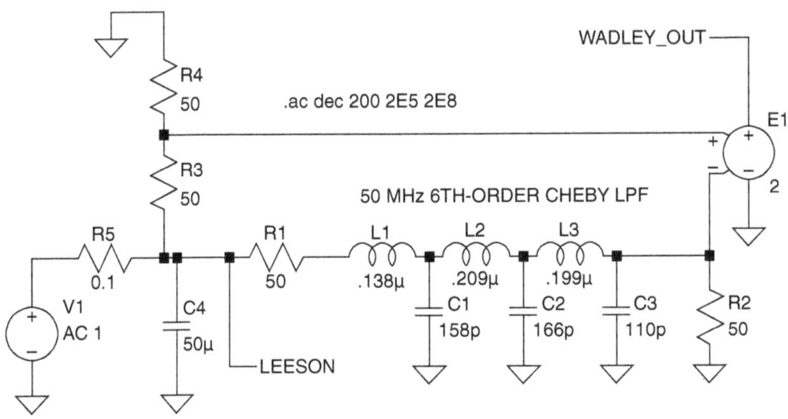

FIGURE 18.2 LTspice model to simulate SSB phase noise profile after a Wadley loop.

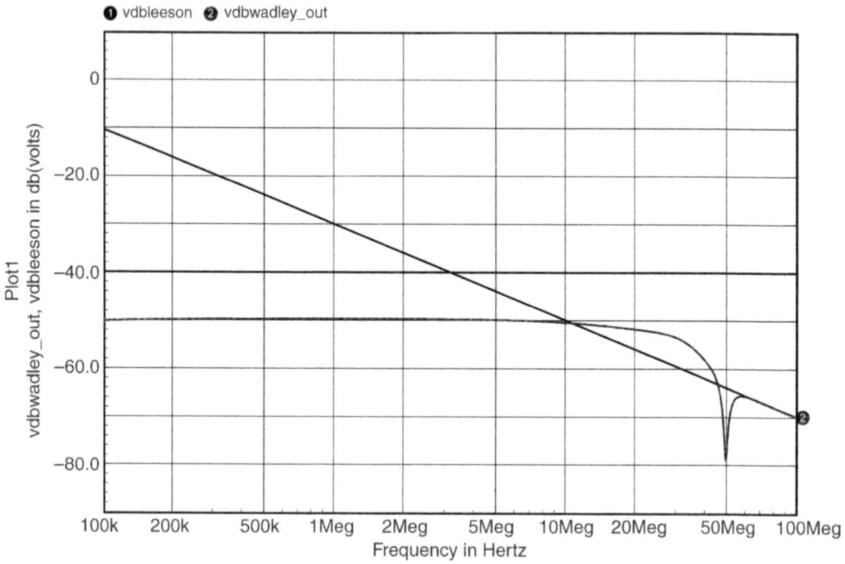

FIGURE 18.3 Phase noise before (sloping trace) and after Wadley loop (SSB) with low-pass filter transformed from RF bandpass.

transform of our heroic filter and consists of R1, R2, L1, L2, L3, C1, C2, and C3. The 50 ohm equal source and load resistors attenuate the signal level by a factor of 2, so we must provide the same loss for the subtractive path using R3 and R4. A VCO phase noise profile is modeled using the RC low-pass filter of R5 and C4 (corner frequency = 32 kHz). Then this LTspice model is simulated as in the file named "SUBTRACTING_FILTER.asc" in the appropriate folder in the toolkit. Running the simulation yields the curves of Figure 18.3. The −6 dB/octave sloping line is the VCO phase noise at a distance from the carrier equal to the frequency in Hz shown on the x-axis. The other curve is the resulting phase noise left unsuppressed by the Wadley loop. Observe that suppression occurs below 10 MHz at a rate of 6 dB/octave, which flattens out the VCO phase noise of −6 dB/octave. At 100 kHz, the VCO noise is attenuated by 40 dB, and at 1 MHz from carrier it is improved by 20 dB. We have not shown the shelf of VCO noise above the Leeson knee (discussed earlier). This VCO phase noise descent of −6 dB/octave from 0 to 100 MHz is typical of a 3 GHz VCO having a resonator Q of 30. In Figure 18.4 , the heroic filter is replaced by a 16 ns delay line. Notice the same shape of noise curve as in the previous figure, except above about 20 MHz, where the standing wave of the delay line equates to a full wavelength at 62.5 MHz, resulting in the notch shown. To simulate the Wadley noise improvement without resorting to a bandpass-to-low-pass transform, we provide the model of Figure 18.5. Notice that the low-pass characteristic is now a 40 MHz bandpass centered at 300 MHz for this example. A 350 Kelvin noise source drives a VCO centered at 300 MHz (with 10 MHz/volt Kv) to generate flat frequency response FM, which is equivalent to a −6dB/octave PM

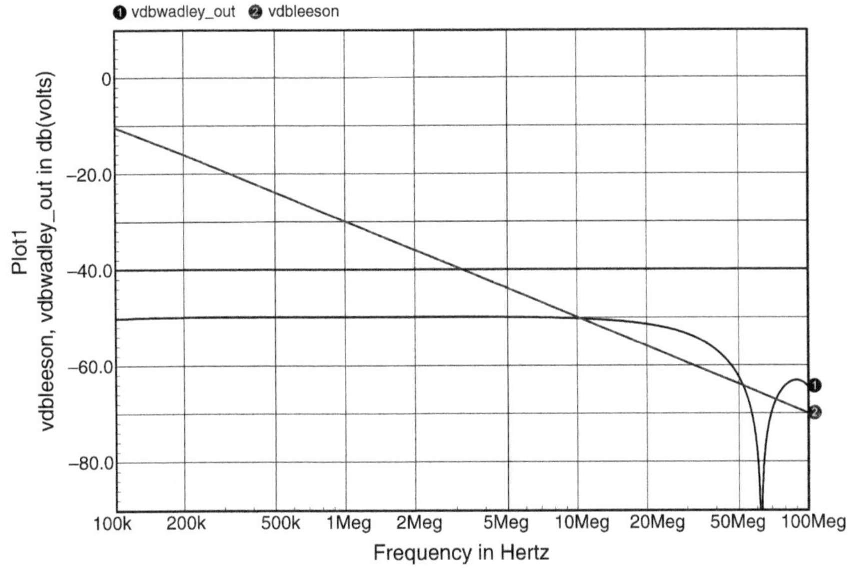

FIGURE 18.4 Phase noise before (sloping trace) and after Wadley loop (SSB) with low-pass filter replaced by an equivalent delay line (at zero Hertz).

noise spectral density (recall that the VCO is an integrator for phase). The RF spectrum is shown in the plot on the left, and the Wadley loop output is shown in the plot on the right. Figure 18.6 is a zoomed outview of both plots of Figure 18.5. Notice the "platform" of noise within the filter passband.

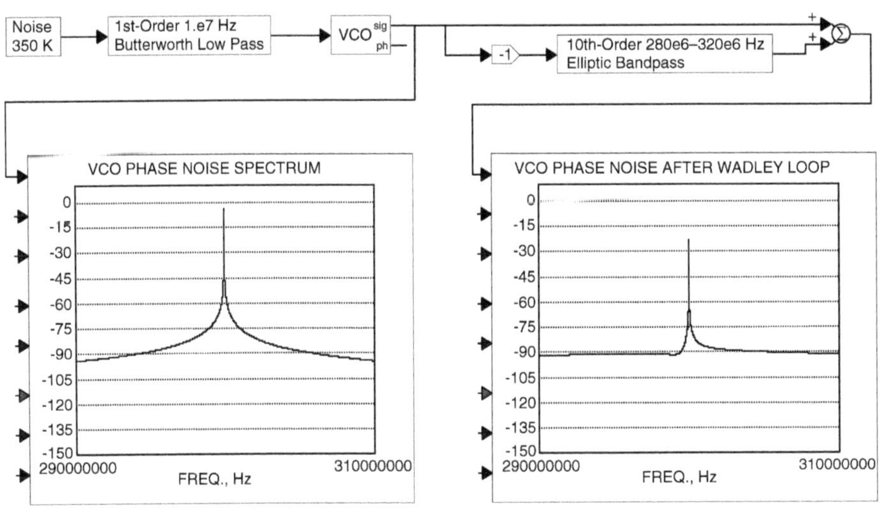

FIGURE 18.5 Phase noise before (left spectrum) and after (right spectrum) Wadley loop at RF (with bandpass filter).

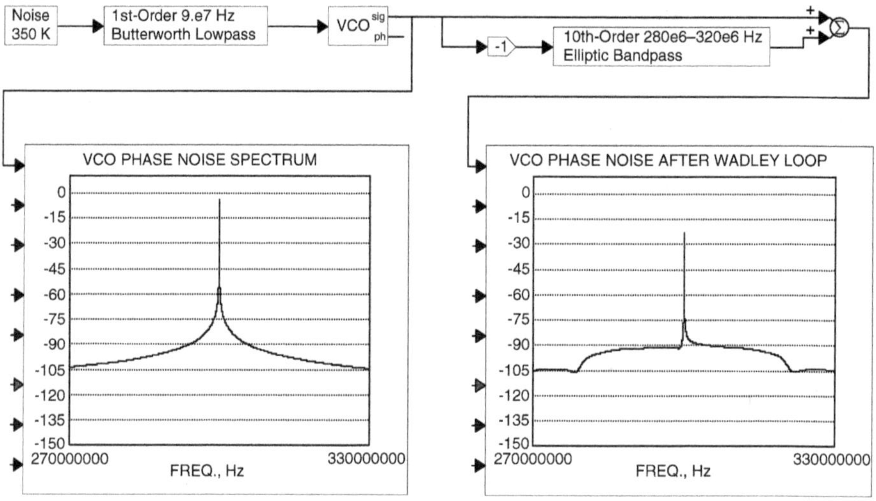

FIGURE 18.6 Zoomed-out view of phase noise curves of Figure 18.5.

In Figure 18.7 , we have changed the LTspice SSB noise model for the heroic filter to a 16 ns delay line and we have also eliminated the VCO sloping noise by removing the 50 uF capacitor (C4) of Figure 18.2 (see LTspice simulation model named "SUBTRACTING_DELAY-LINE.asc"). We can now see the effect of the Wadley loop on a flat PM spectral density in Figure 18.8. Since a 16 ns delay line is a full wavelength long at 62.5 MHz, it is a quarter wavelength long at 15.625 MHz, or 90° in delay phase. Two equal amplitude quadrature signals at 15.625 MHz will produce a bump (gain) of 3 dB at 15.625 MHz, and indeed the figure shows that. At half this frequency (7.8125 MHz), only 45° lag is experienced in the delayed path. This yields a level of 20 $\log_{10}(0.7654)$, or -2.32 dB at 7.8125 MHz, and -6 dB/octave from there as frequency decreases.

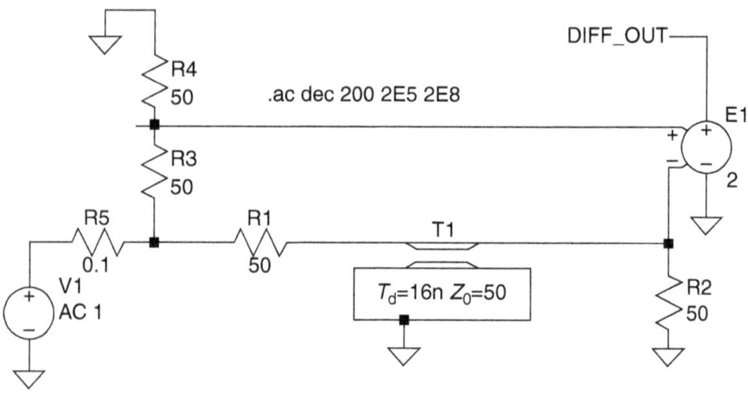

FIGURE 18.7 Simulation model for Wadley loop without Leeson's phase noise slope, so that effect of filter delay can be observed.

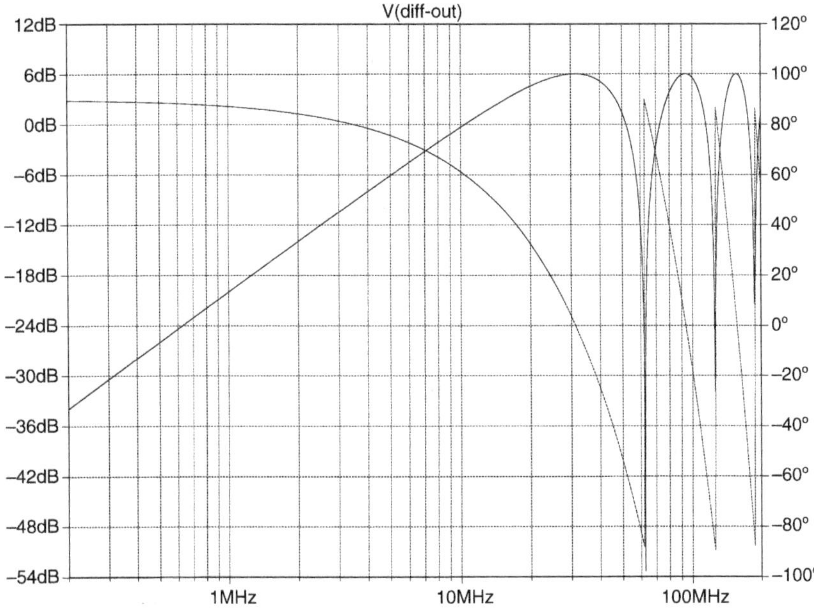

FIGURE 18.8 Frequency response of Wadley loop alone, with delay line to represent the filter (SSB); magnitude response peaks at 6 dB; phase response inflects at delay line wavelength multiples.

Thus, we see that the Wadley loop improves VCO phase noise sidebands at 6 dB/octave as we approach the carrier frequency from either sideband's direction. The same improvement is applicable to close-in spurs. To move the Wadley noise corner upward, and thereby obtain more noise suppression, we need to widen the filter, or reduce the number of sections, either of which has the drawback of allowing more of an adjacent 50 MHz comb sideband to enter the Wadley loop. Typically, the heroic filter in the Wadley loop will be a cavity filter for its high Q (center frequency divided by passband width). At low frequencies (under 1 GHz), a multisection cavity filter becomes physically large. The heroic filter can be made simple and nonheroic by replacing the comb generator with a narrow-spectrum synthesized signal, although noise performance likely will be poorer than the comb generator.

It may be desirable to reduce the maximum harmonic number N of the comb generator, and this can be accomplished by recentering the frequency ranges.

One additional advantage of the Wadley loop is its fast frequency switching and settling speed, making it attractive to pair with an analog direct synthesis architecture.

Direct analog frequency synthesis needs no PLL, and exhibits fast switching from one frequency to another. Recursive mixing and dividing enable very fine frequency resolution, which can be made modular (increasing the number of essentially identical modules improves the frequency resolution by a decade per module if based on decade synthesis). One decade synthesis technique is the biquinary approach used in the PTS (Programmed Test Sources®) synthesizers such as model

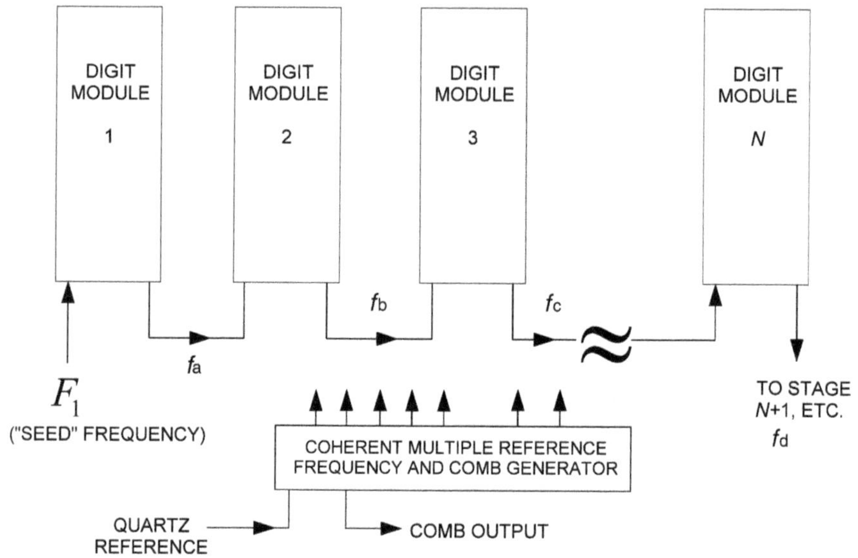

FIGURE 18.9 Block diagram of direct analog frequency synthesis (final output may be heterodyned to a higher starting frequency; model for such a heterodyne stage is not shown).

PTS-160. The block diagram is shown in Figure 18.9. The frequency resolution improves by a decade after each subsequent digit module. The frequencies at intermediate points in the figure are

$$f_A = F_1 + \alpha_1$$

$$f_B = F_1 + \alpha_2 + \frac{\alpha_1}{10}$$

$$f_C = F_1 + \alpha_3 + \frac{\alpha_2}{10} + \frac{\alpha_1}{100}$$

$$f_D = F_1 + \alpha_N + \frac{\alpha_{N-1}}{10} + \frac{\alpha_{N-2}}{100} + \cdots$$

Figure 18.10 shows the detailed block diagram of an individual analog synthesis "digit module." In the PTS-160 synthesizer, $\Delta = 1$ MHz and $F_1 = 14$ MHzand thus the one-of-two selector chooses either 112 or 113 MHz, and the one-of-five selector chooses 14 or 16 or 18 or 20 or 22 MHz. The result after four digit modules is a frequency from 14 to ~15 MHz with a step size of 100 Hz. After seven modules, the resolution is 0.1 Hz. The recursion allowable depends only on physical space and power available. Each progressive module divides its output by 10, so spurs from the previous module are reduced 20 dB. The final module fairly dominates the spur floor amplitude. Direct analog frequency synthesis can, therefore, be theoretically very high in performance. The PTS-160 combines direct synthesis with a Wadley loop to expand

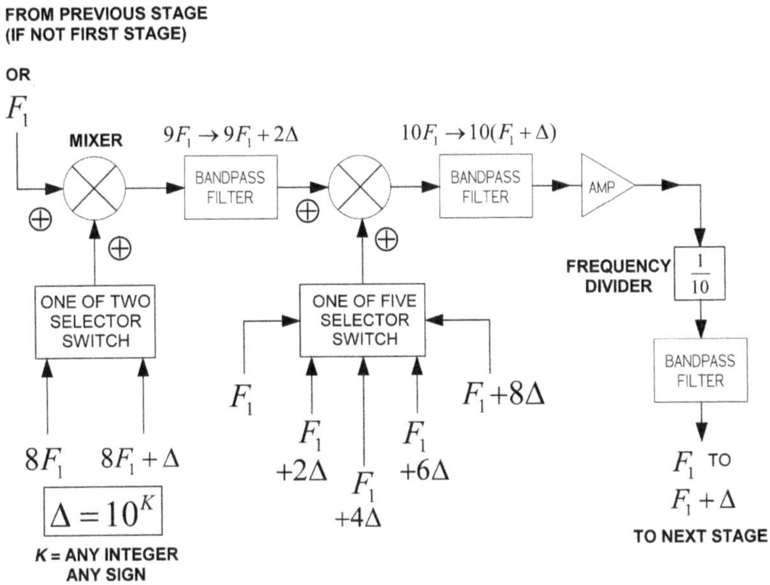

FIGURE 18.10 Block diagram of a digit module in a cascade of similar modules in a direct analog frequency synthesizer.

the frequency span to 10 MHz increments and final coverage is 100 kHz–160 MHz (−0.1 Hz in a fully populated chassis).

18.2 INDIRECT FREQUENCY SYNTHESIS WITH PLLs

A PLL can be programmed to yield a frequency agile output. Such an agile PLL example is shown in Figure 18.11 using the Analog Devices® chip ADF4150HV, capable of either integer or fractional-*N* synthesis. A VCO from ZCOMM® is used to generate an output frequency between 400 and 450 MHz, so that the resulting PLL can be applied to the Wadley loop example previously described in place of the direct analog synthesis. In frac-*N* mode, this chip can produce fair phase noise performance. Figure 18.12 shows the total phase noise (top curve in the appropriate set) for the frac-*N* modulus of 500, with a clock of 10 MHz, phase detector frequency of 10 MHz, and frequency resolution of 20 kHz. These curves were obtained using the free Analog Devices® software "ADIsimPLL."

This chip can also be programmed to operate as an integer-*N* PLL. See Figure 18.13. This constrains the loop feedback counter to integer values only, and thus the phase detector must operate at the same frequency as the desired resolution, in our case 20 kHz. Thus, at 400 MHz, the divider noise penalty is $20\ \log_{10}(4E8/2E4) = 86$ dB. This explains why the output phase noise curve near-zero offset frequency is so elevated (see Fig. 18.14) compared to the frac-*N* mode. But even worse, the spurs due to the phase detector sampling frequency are huge for

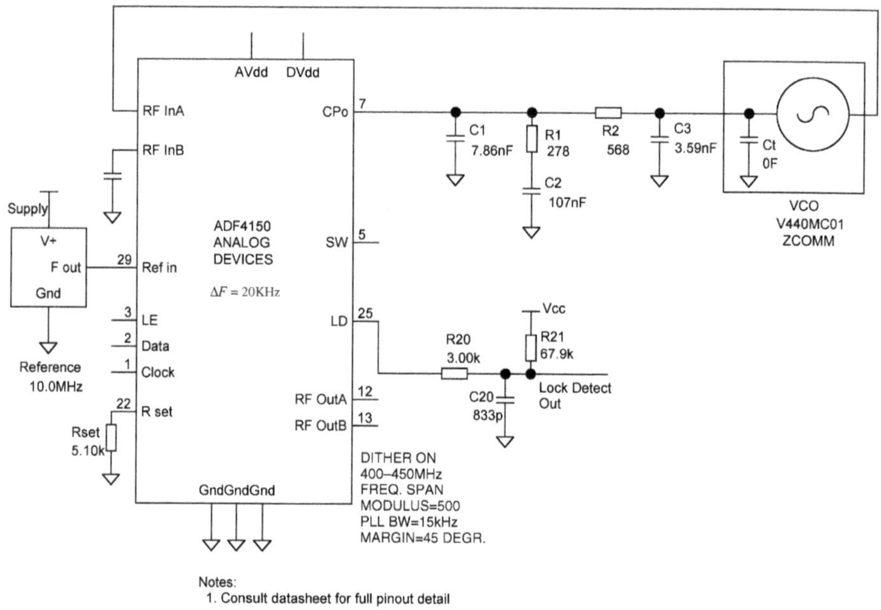

FIGURE 18.11 Topology for fractional-*N* PLL synthesizer for 400–450 MHz range having fair phase noise performance, using a commercial chip.

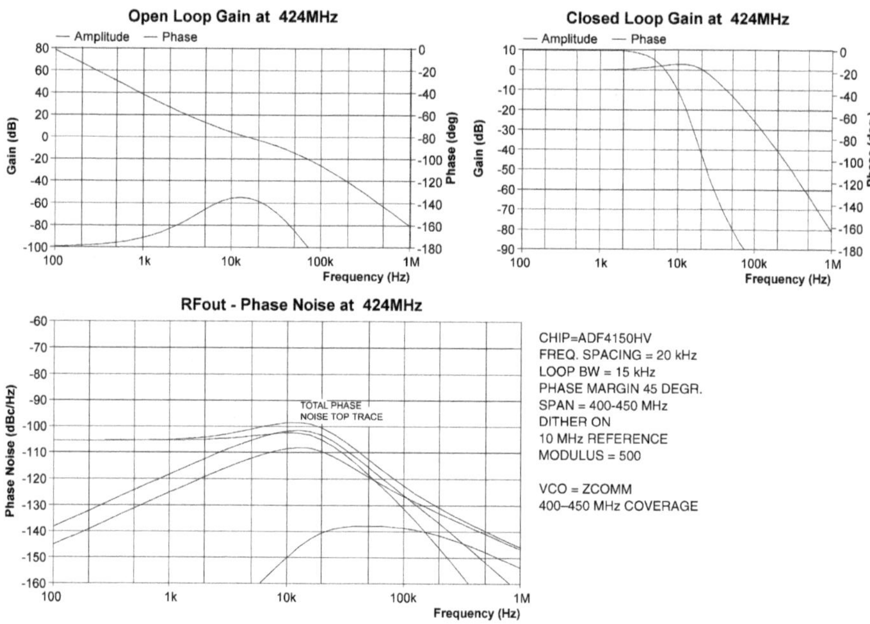

FIGURE 18.12 Some performance curves for the frac-*N* PLL of Figure 18.11.

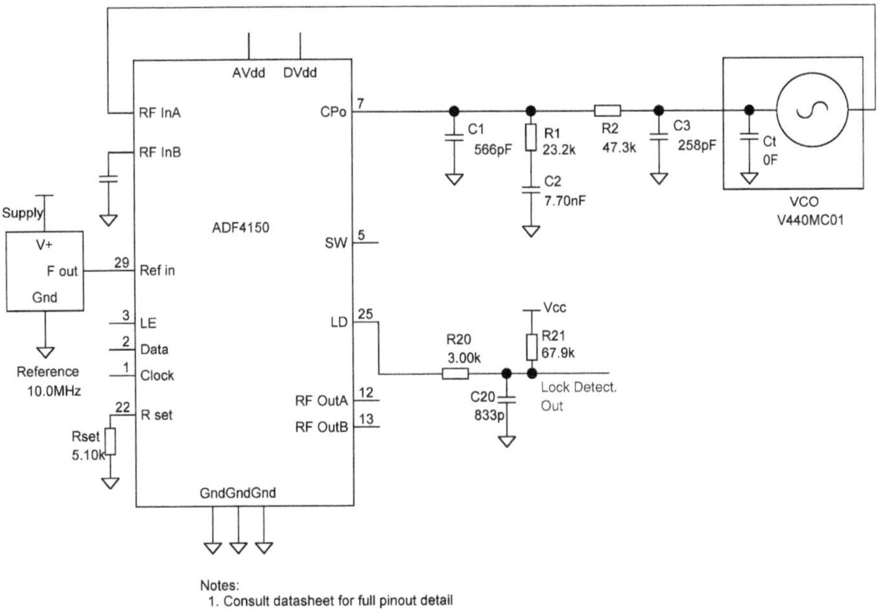

FIGURE 18.13 Topology for an integer-*N* PLL synthesizer for 400–450 MHz range having poor phase noise performance and charge pump leakage (unbalance) caused sampling spurs using a commercial chip.

the case of even a tiny charge pump leakage current. These can be nulled out by applying a tiny DC current from a trimpot and resistor network, tweaked to null these sidebands. However, they may remain nulled only across a narrow temperature range. Logistically that can resemble trying to balance a pyramid on its point or trying to stand up in a hammock. Enter the heroic filter approach that is difficult or impossible to implement due to the one-port high-impedance nature of the charge pump.

As we have just shown, a PLL synthesizer can be employed together with a Wadley loop, but some of the Wadley loop phase noise benefits may be wasted.

18.3 SIMPLE FREQUENCY ACQUISITION IMPROVEMENT FOR A PLL

A benefit of integer-*N* synthesis is the greatly widened capture range caused by frequency division. For example, a 1 kHz PLL bandwidth paired with a feedback counter of unity division factor will have only a capture range of about 1 kHz for minimal cycle slips. But a 1 kHz bandwidth loop using a feedback counterdivision factor of 1000, although poorer in phase noise performance by 60 dB near-zero Hz offset, will have a capture range of almost 1 MHz at the VCO output. Thus, feedback division factor (along with the necessary reference division factor) can be used as a

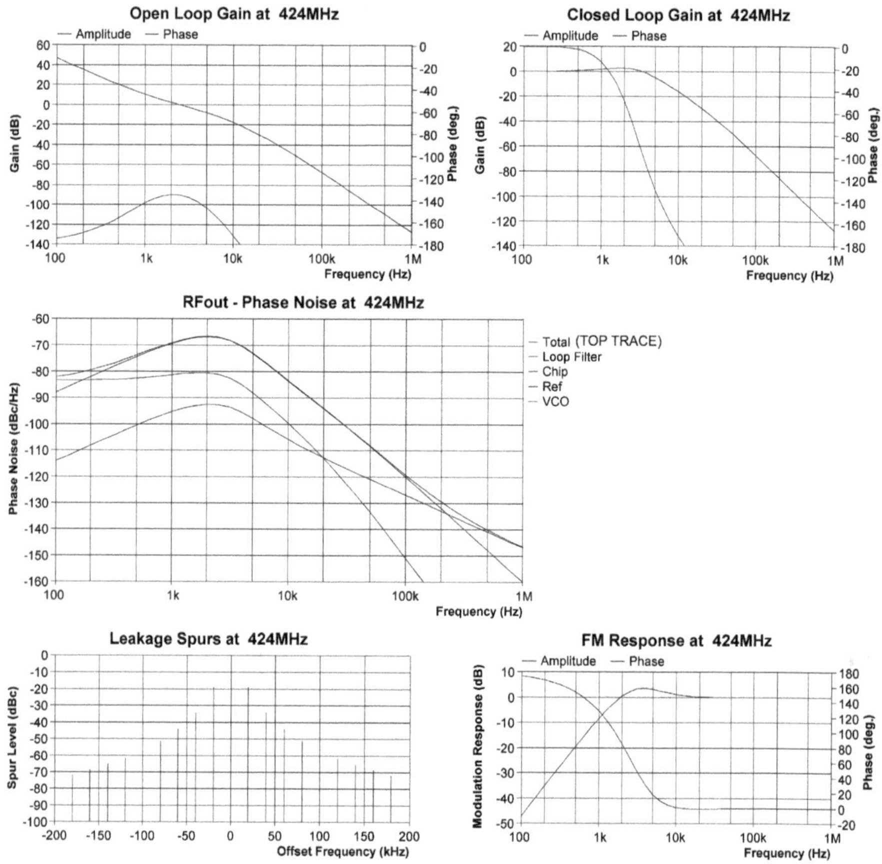

FIGURE 18.14 Performance curves for integer-*N* PLL synthesizer of Figure 18.13 (note high spur levels also due to phase detector offset or charge pump leakage).

frequency acquisition tool, at the expense of phase noise. In some applications, the resulting pedestrian performance is more than adequate. Figure 18.15 shows the drastic improvement realized by inserting a frequency division by 10 in both reference and VCO paths (and recalculating loop filter parameters to keep the same natural frequency and damping).

This is a powerful technique that can be used in many PLL applications (but not in instances where pulses or edges might be occasionally missing). Although the simulation here uses analog phase detectors, the argument is valid for other types as well, including the phase frequency detectors (PFD) although quantitative results may differ.

18.4 HYBRID FREQUENCY SYNTHESIS WITH DDS AND PLL

A PLL and a DDS can be combined in several ways. A PLL can be used for the higher digits of resolution, together with a DDS heterodyned to yield the finer resolution

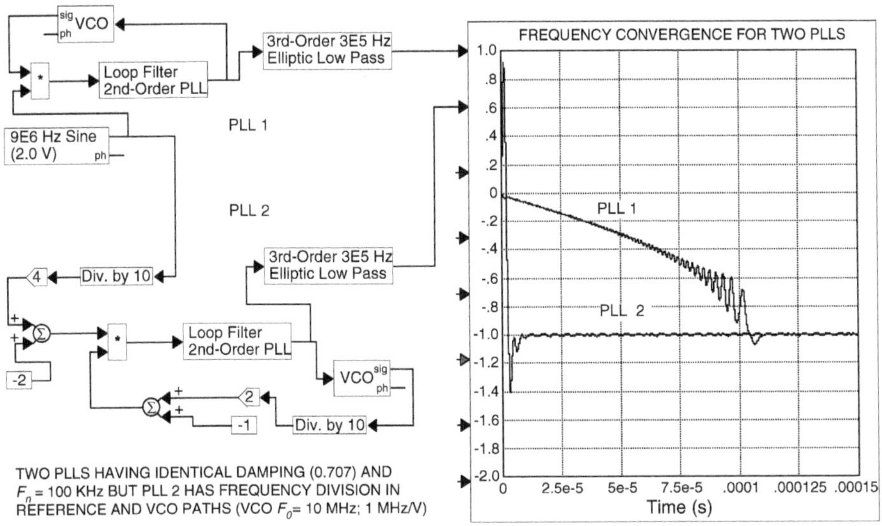

FIGURE 18.15 Simulation showing the frequency acquisition speed benefit of including digital dividers (counters) in reference and VCO paths.

digits. Or, the DDS can be used as the feedback counter in a conventional PLL, yielding many of the performance benefits of a frac-N PLL without the need for dither and with the benefit of very high resolution. Figure 18.16 shows such a configuration. Normally, the output frequency of a DDS is given as

$$f_{\text{out,dds}} = \frac{M}{2^n} \cdot f_{\text{clock}} \tag{35}$$

where M is code word value (decimal equivalent), n is number of bits available for full scale, $f_{\text{out,dds}}$ is output frequency in Hertz of DDS DAC, and f_{clock} is DDS clock frequency in Hertz.

Since the DDS is used as the PLL feedback element, the equation for the PLL output frequency is

$$F_{\text{out}} = F_{\text{in}} \cdot \frac{2^n}{M} \tag{36}$$

For avoidance of closely adjacent images, $F_{\text{out}} > 2F_{\text{in}}$.

Let us consider an example. Suppose, we have a 32-bit frequency code ($n = 32$). Furthermore, suppose we have a 10 MHz reference frequency F_{in} and we want a 200.01 MHz output frequency F_{out}. Then, solving for the code word yields $M = 214737628.1$ that equals CCCA2DC (in hex format) to the nearest round integer. The actual frequency will then be 200.009999924186552 MHz, an error of -0.0758 Hz.

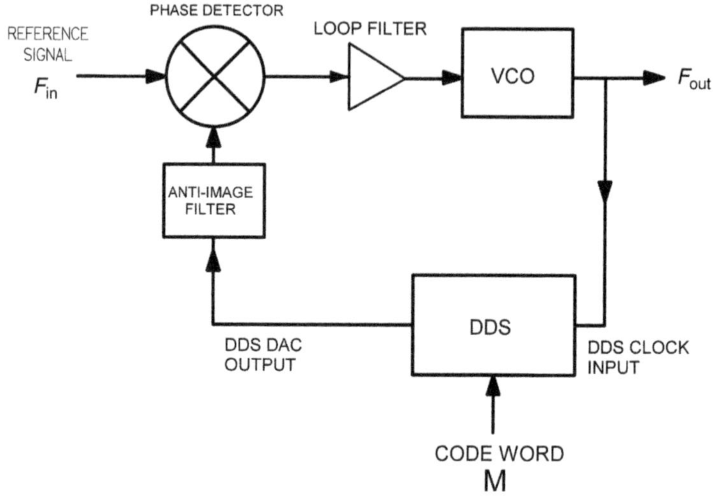

FIGURE 18.16 DDS employed in PLL feedback path to synthesize a high-resolution signal frequency.

For example, if $F_{in} = \{100/10.24\}$ MHz, and we wanted $F_{out} = 200.0$ MHz, the code word would be 209715200 exactly and the hex value would be C800000 and the frequency error is zero. If we increment the code value by 1, we get C800001 = decimal 209715201 that yields a new frequency of 199.99999905 MHz. A DDS capable of clocking at > 200 MHz would be needed. For the above examples, generating 200 MHz would be accompanied by image spurs at the phase detector at approximately 190 and 210 MHz. But generating 15 MHz would produce images at 5 and 25 MHz that is in the phase detector operating range and would not be easily ignored. Therefore, this configuration should be limited to $F_{out} \gg 2F_{in}$.

An interesting observation is that by sweeping the digital code word, the PLL can be equivalently swept for acquisition, or a PFD can be used for wide capture. However, sweep range should be limited such that the antiimage filter output does not enter its stopband. Such a problem might result in the phase detector output stalling and the loop latching up, especially if a PFD is used.

Another way to combine a PLL and a DDS is to use the DDS to generate a variable reference frequency. This can be combined with *N*-over-*M* synthesis to produce a nearly exact desired final frequency without frac-*N* techniques. Of course, the noise and spurs of the DDS will be elevated by the PLL feedback factor *N* if they fall within or near the PLL passband.

Yet another way to combine a PLL and a DDS is to use the DDS as a fine-resolution sideband on the coarse-resolution PLL. See Figure 18.17. A DDS feeds a single-sideband mixer (to reduce the burden on the subsequent bandpass filter if desired) and shifts the VCO output frequency prior to dividing by *N*. Many DDS chips have a 90° quadrature pair of outputs, so the respective splitter is inherent. A sample of the VCO is split into quadrature waveforms and applied to the same

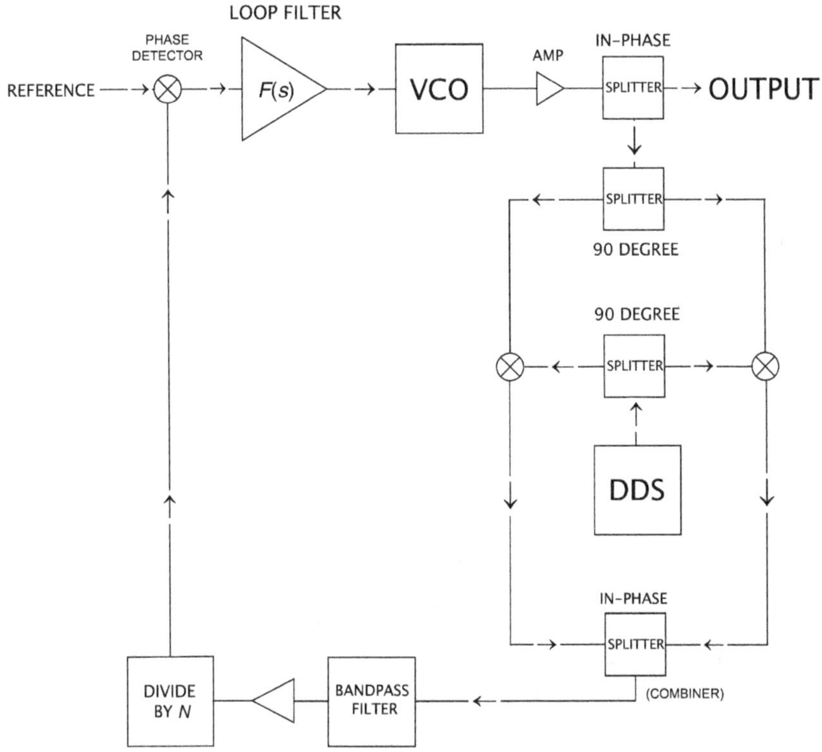

FIGURE 18.17 A DDS-produced improvement to the resolution of a PLL synthesizer.

mixer pair fed by the DDS quadrature signals. The result, when summed, is a frequency-shifted signal that is offset from the VCO frequency by the DDS output frequency.

Assume an example wherein the reference frequency is 10 MHz. Then programming the feedback divider to any integer value results in 10 MHz output frequency steps. The 10 MHz gaps are then filled in by the DDS, which can typically produce steps as fine as a fraction of a Hertz. Moreover, this entire scheme can be augmented with a Wadley loop to provide wider span with larger steps. It is important to buffer or isolate the VCO from not only the user's load but also from the limited reverse isolation of the mixers and splitters. If this is not done, the DDS offset sideband may leak through to the output. The loop filter is responsible for cleaning up sidebands inside the PLL, both from the reference leakage and from the sidebands from the DDS path (because the SSB mixer isn't perfect).

This scheme can use any type of phase detector, but if a PFD is not used, sweep may be required for acquisition. This type of synthesis approach yields fairly low phase noise and good spur performance outside the PLL bandwidth. Most PLL chips (such as the ADF4150) can be used for the PLL section, and many DDS chips with built-in quadrature outputs are suited to this application. The block diagram does

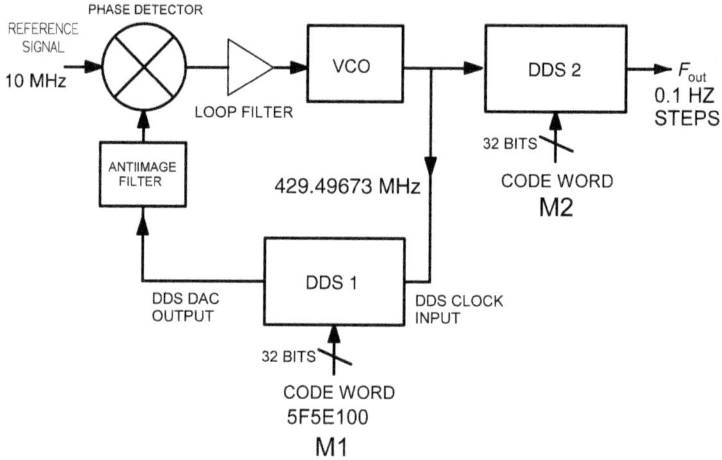

FIGURE 18.18 Forcing a DDS 2 to generate decimal-friendly output frequencies by adding a second DDS 1 and a PLL.

not show the clock source for the DDS, but many chips have built-in frequency multipliers that can generate the clock from the common reference signal (10 MHz in our example). A DDS feedback PLL can be used to force the sideband DDS steps to be decimal-valued, rather than nondecade. For example, suppose we want a 32-bit DDS to have decimal steps, we first divide its clock frequency (example, 500 MHz) by 2^{32}. The result will not be pretty for a 500 MHz clock. But if we use a new clock frequency of 429.49673 MHz, the DDS step size will be exactly 0.1 Hz. To match a 10 MHz reference clock, we need 10^8 steps, or 5F5E100 in hex. This situation is shown in Figure 18.18. DDS 1 in the PLL feedback path, programmed for M1 = 1E8 (5F5E100 HEX), establishes 10 MHz to match the 10 MHz reference at the phase detector. Its clock now sits at 429.49673 MHz and is used by DDS 2 to provide it with 0.1 Hz frequency steps multiplied by code word M2. We now have a decimal-friendly DDS 2.

Why do we care about tiny frequency errors of milli- or micro-Hertz? Consider a signal of 200 MHz whose jitter we wish to observe. Also consider that this signal is synchronized to a 10 MHz jitter-free clock that is available. We wish to synchronize our synthesizer to the same 10 MHz reference, and set our synthesizer to 200 MHz so that we can trigger an oscilloscope in a jitter-free fashion. If our 200 MHz is off by 128 milli-Hertz due to DDS messy fractions, then every 7.8125 s our oscilloscope trace will drift horizontally by 360° of phase while trying to observe the jitter on the incoming jittery 200 MHz signal. In other words, there are situations in which slaving two different synthesizers to a common reference should synchronize any signal generated by one or the other without frequency drift or phase creep if the "knobs" are set to expect such coherence. If two synthesizers are slaved to a common reference, and the "knobs" are set for harmonic ratio of one to the other, one should be able to expect exactness of the said ratio.

For more tutorials on DDS, see Refs [32,33].

18.5 PHASE NOISE CONSIDERATIONS

How does one model the phase noise of the VCO and the closed loop, including reference phase noise multiplication? We will demonstrate how to perform such modeling using LTspice.

Recall the Leeson curve for phase noise of an oscillator. There is a knee at $f_{osc}/2Q$ below which noise rises at 6 dB/octave (20 dB per decade). Somewhere below this knee is a $1/f$ knee and below this $1/f$ knee noise rises 9 dB/octave (30 dB per decade). In most PLL applications, it is desired to shrink most of the VCO noise below the $f_{osc}/2Q$ knee. The PLL shrinks the said noise by 12 dB/octave, so most VCO noise gets "eaten," even the $1/f$ noise. Thus, it is normally adequate to consider only the normal Leeson noise excluding the $1/f$ component. ($1/f$ refers to power, not voltage, thus it alone rises 3 dB per octave, and is superimposed on the Leeson noise of 6 dB/oct, making the total 9 dB/octave)

Figure 18.19 shows how Leeson noise can be modeled using LTspice. R1 is used as a perfect noise source (a 1 k ohm resistor generates 4 nV per root Hertz of noise, so to generate a −120 dB noise level requires a 63 megaohm resistor). At high frequencies, the gain of E1 is unity (the ratio of R3 to R2). Amplifier E2 is set to unity gain if the user wants −120 dBc of SSB phase noise shelf above the Leeson knee. The corner of the shelf (the $f_{osc}/2Q$ knee) is set by $1/(2\pi C2)$ thus for $C2 = 1.6$ uF, the knee occurs at 100 kHz. By making the gain of E2 equal to 0.01, the shelf will be at −160 dBc, and by setting E2 to 10, the shelf will be at −100 dBc. Figure 18.20 shows the phase noise versus offset frequency for a shelf of −120 dBc. Notice how perfectly it conforms to the Leeson curve for a 10 MHz oscillator using a tank Q of 50, for example.

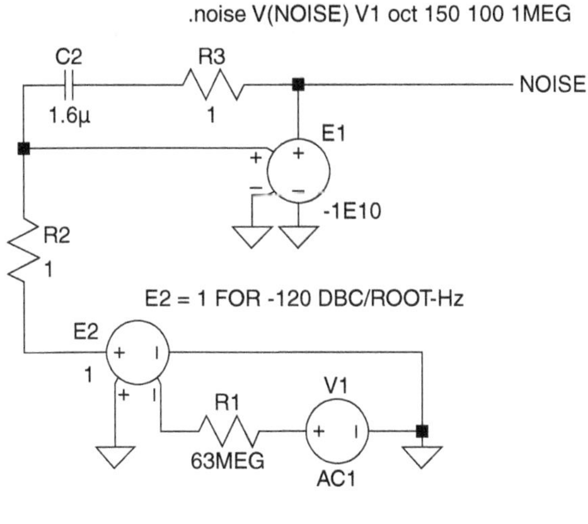

FIGURE 18.19 LTspice model for simulating a classical oscillator's Leeson phase noise curve.

FIGURE 18.20 Resulting phase noise plot for the model of Figure 18.19.

The Leeson noise model for LTspice (see the file named "VCO_PHNOISE.asc" in the toolkit in the Spice files folder) generates noise whose units are dBc/root-Hz. We can pretend, however, that the units are dBc/Hz provided that we maintain consistency throughout our simulation, and treat any calculations as voltages, not power. To integrate the noise between any two frequency boundaries, we must display and plot ONOISE, not 20 $\log_{10}$ONOISE. Then, we hold down the control key while clicking the left mouse button over the label ONOISE on the plot. A box will appear with the resulting integrated value in volts RMS, which we will treat as the square of the phase jitter. Double the value of the phase jitter (after calculation of square root) for radians of phase jitter, since DSB jitter is twice SSB jitter (they are 100% correlated sidebands, not statistically independent, so voltage addition is appropriate rather than power addition).

Next, let us incorporate the VCO phase noise model of Figure 18.19 into the closed loop of a PLL whose natural frequency is 100 kHz, and we get the simulation model of Figure 18.21. Amplifier E2 sets the noise shelf at −120 dBc, and C3 sets the noise corner at 100 kHz. The LTspice file is named "Type-II_PLL_VCO_PHASE_NOISE_INCLUDED_Response.asc" and appears in the Spice files folder of the toolkit should the reader wish to run the simulation and "kick the tires." Notice in the resulting phase noise plot of Figure 18.22 how the VCO shelf noise of −120 dBc is maintained above 100 kHz, but the VCO climbing noise below 100 kHz is eaten by the PLL by 12 dB/octave, for a net noise residue of −6 dB/octave as frequency decreases below 100 kHz. Of course, the reader can change the PLL bandwidth and damping by simply rerunning the toolkit file "PLL_Synth.exe." Note that the op-amp

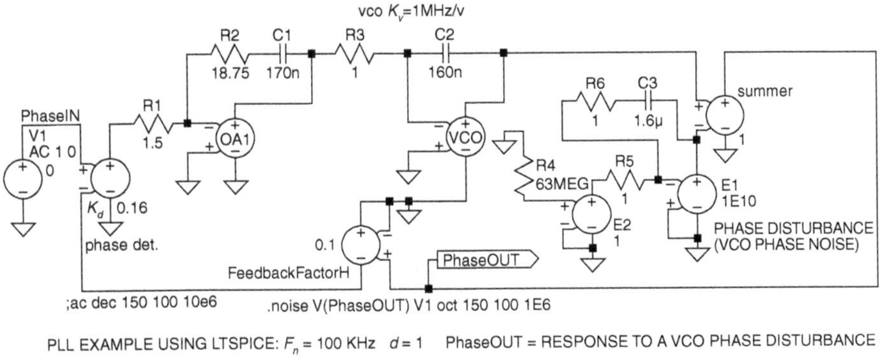

FIGURE 18.21 Example of closed loop PLL including VCO noise model.

input resistor R1 and the damping resistor have been scaled downward by a factor of 1000 to minimize their noise contribution for this simulation, but they must be scaled upwards in practice. Doing so will lift the −160 dBc shelf near 100 Hz due to resistor noise in the loop filter (or op-amp). Op-amp loop integrator capacitor C1 was also scaled upward by 1000 to preserve the R2C1 time constant.

What about the noise contribution of the reference signal? We can apply the Leeson noise model of Figure 18.19 to the reference signal as well, resulting in

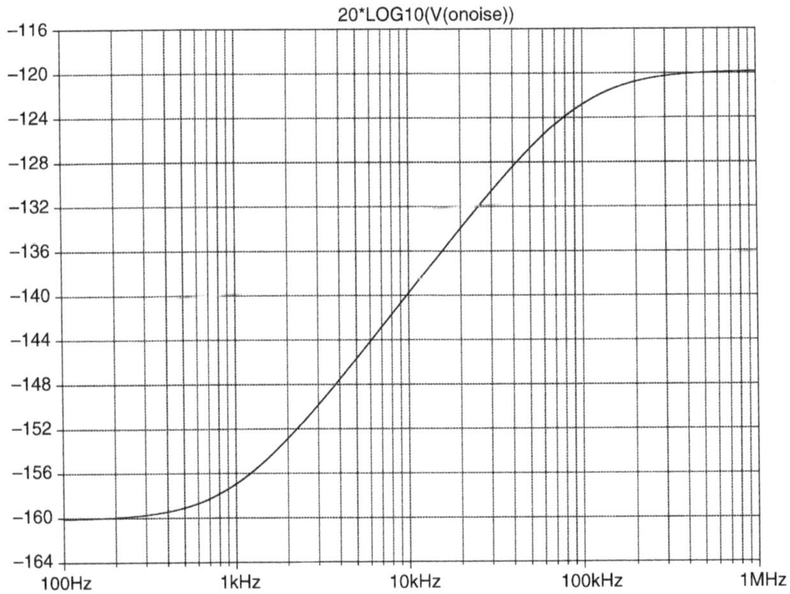

FIGURE 18.22 Resulting phase noise plot for the model of Figure 18.21.

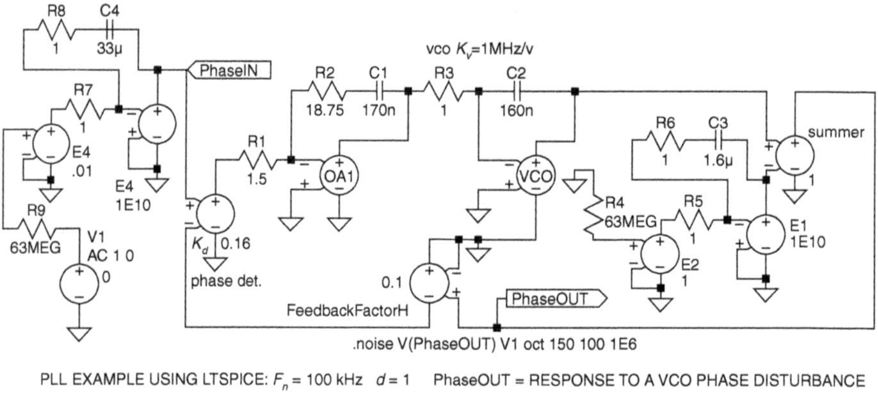

FIGURE 18.23 Example of closed loop PLL including both reference and VCO noise models.

Figure 18.23. Here, we have modeled a crystal oscillator (XCO) of −160 dBc/Hz SSB noise shelf (since E4 = 0.01) and a 4.82 kHz Leeson knee. This would be typical of an XCO having a loaded crystal Q of 1030 (a mediocre XCO at that!). The phase noise of the XCO reference gets elevated by 20 dB because the PLL feedback frequency division is 10 (FeedbackFactorH) and since $20\log_{10}10 = 20$. Thus, we see

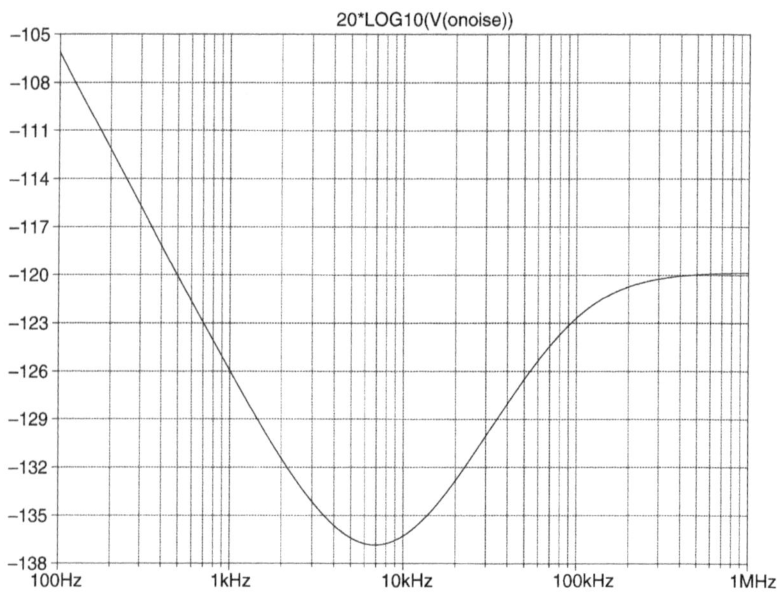

FIGURE 18.24 Resulting phase noise plot for the model of Figure 18.23.

about −140 dBc near the 10 kHz region (see Fig. 18.24) instead of the −160 dBc XCO noise shelf (10 kHz is above the XCO noise corner and well inside the PLL passband). An actual well-designed PLL synthesizer would probably exhibit a differently shaped noise plot. To "kick the tires," run the LTspice file named "Type-II_PLL_VCO_AND_REFERENCE_NOISE_INCLUDED_Response.asc" in the Spice files folder.

18.6 PROS AND CONS OF DDS-AUGMENTED SYNTHESIS

Every application is potentially unique but there are common attributes of DDS-plus-PLL and DDS-plus-direct-analog synthesis designs. As pointed out in the previous section, if a DDS reference clock has a frequency that is decimal-friendly (e.g. 10 MHz or 200 MHz or 500 MHz), the DDS output frequency will be decimal-hostile since it is a multiple of the DDS granularity (resolution) that will be decimal-hostile (clock frequency times 2^{-N} where N is number of frequency control bits). In order to output decimal-friendly frequency increments, the DDS clock must be specially derived. This necessitates two DDS.

Another disadvantage of synthesizers that include DDS is the relatively high spur floor density. Spurs are small taken individually, but collectively appear very closely spaced in frequency and thus can be problematic. Spurs add as 20 log N for a small number of equal-amplitude spurs within a slice of passband (thus 10 equal-amplitude spurs look 10 times as large as one for their contribution to jitter). This is also true in general for communications systems.

Another disadvantage of DDS is the need for heroic output filtering, since the raw output is staircased. Even with a large degree of smoothing, these staircase steps tend to "crawl" like escalator steps along the slopes of the output sine wave, so that if one attempts to feed the filtered output into a zero-crossing comparator to form a square wave, the square wave edges will snap back and forth in the time domain by a small amount as the escalator steps climb the waveform. This problem is reduced by increasing the number of stairsteps (increasing the phase accumulator resolution) or increasing the selectivity of the output filter, but is never eliminated.

Advantages of DDS-augmented synthesizers include (a) fine frequency steps, (b) moderately low phase noise if spurs are not included, (c) fast frequency switching, and (d) ease of application, since the DDS (minus filter) is often available as a chip (IC).

18.7 MULTIPLE LOOPS

Another way to improve synthesizer resolution is by using multiple PLL loops. The approach is similar to the DDS sideband method, except another PLL or multiple PLLs are substituted for the DDS. Each successive PLL has a finer resolution than the previous one, usually by 10 times. Its frequency translation spurs are absorbed by its previous PLL's selectivity just as in the DDS sideband offset method previously discussed.

18.8 REFERENCE SIGNAL CONSIDERATIONS AND FILTERING

The reference signal for a frequency synthesizer is probably the most important component for noise reasons. That is because most synthesizers multiply the reference frequency and hence the amplitude of its noise sidebands. Of course, other components are also contributors, but in most cases can be made minor. The VCO noise, for example, can be reduced by widening the PLL bandwidth (until the divider noise becomes dominant). The phase detector noise can be minimized by using analog technology. Edge triggering noise aliasing can be eliminated by using analog phase detection and analog frequency division in the final PLL feedback counterstage.

Reference signal amplitude is also pivotal. The slope of the zero crossings of the reference signal compared to the phase detector input noise determines jitter and hence phase noise of the interface. One way to raise this SNR is to make the reference signal amplitude very large and then bandpass it to limit noise power from it, and apply this large amplitude waveform to a passive clipper using Schottky diodes. For example, use a 10 V reference amplitude (which would burn out the stage it feeds) but clip it at +/−0.8 V. The slope as it crosses zero volts will now be the same as that of the 10 V signal. Such an example is shown in Figure 18.25. Use diodes having fast response and low 1/*f* noise. A "sow's ear" crystal oscillator reference can be made a "silk purse" for near-carrier phase noise by adding the "poor man's crystal bandpass filter" comprised of a 10 MHz series resonant crystal driving a 100 ohm load (R1). This filter then drives an amplifier and then a coarse bandpass filter ("roofing filter") to limit noise harmonics, and finally a pair of back-to-back diodes that clip the signal amplitude to a safe level to drive the phase detector, in this case a digital edge-triggered type (PFD). This technique (with some details that are not shown) can make a marked improvement in phase noise performance in synthesizers employing digital PFDs.

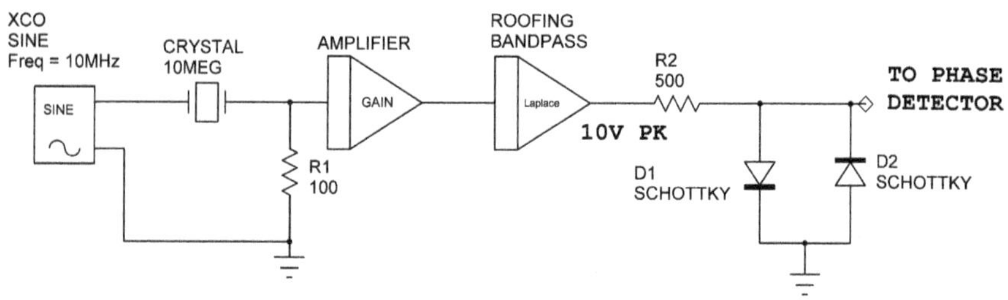

FIGURE 18.25 Technique for raising the SNR of a reference signal interface as seen by a digital phase detector.

18.9 SNR OF VARIOUS PHASE DETECTORS

The phase noise floor of various types and brands of phase detector can be quite different from one another. We have already shown that edge-triggered phase detectors, such as the PFD, suffer from 10 $\log_{10} f$ degradation in noise as their operating frequency increases.

In addition, even passive phase detectors often exhibit $1/f$ noise. But for most conditions, the phase noise is fairly flat versus offset frequency if $1/f$ noise is excluded. See Figure 9.2 for the phase noise of the Hittite process for 100 and 1280 MHz operating frequencies.

All of the papers written that evaluated the phase noise measurements of various types of phase detectors found the diode ring mixer to be the quietest compared to all digital types. Improving the noise level of charge pumps is largely a fool's errand, since the major factor for the poor noise behavior is simply the impulse sampling (edge-triggered operation) as discussed previously in Chapter 9.

GaAs-based devices are poorer by about 20 dB than silicon (which measure −145 to −165 dBc/Hz at nominal 10 MHz operating rate) probably due to $1/f$ noise; above 100 kHz offset frequencies, they converge to similar performance. For the National Semiconductor LMX23XX family, the noise at 10 MHz is −141 dBc/Hz (their published number is −211 dBc/Hz at a 1 Hz sampling frequency; 10 MHz = 10^7 Hz, and 10 $\log_{10} f$ yields a 70 dB higher noise for 10 MHz operation). Divider output noise is generally comparable to the phase detector noise if the phase detector and divider are fabricated with the same core logic circuitry.

RF mixers generally exhibit −170 dBc/Hz performance or better (depending on operating signal amplitudes). Johnson noise (for zero dB noise figure) is −174 dBm/Hz at room temperature. Thus, an operating level of 10 dBm would be expected to yield a −184 dBc/Hz SSB phase noise floor, much superior to digital phase detectors, and this number holds true for almost any operating frequency, unlike edge-triggered devices that degrade at 10 $\log_{10} f$ due to noise aliasing and threshold jitter as we discussed in Chapter 9.

18.10 PHASE DETECTOR DEAD BAND (DEAD ZONE) AND REMEDIATION

This section deals with a unique issue to digital PFD types of phase detector. Due to the flip-flops in a PFD, there can be a race condition when the pulse edges feeding the set and reset nodes occur concurrently. This results in a dead zone near zero phase angle for the type IV (4) PFD (the one that generates minimum output energy when the PLL is locked, illustrated in Figure 2.7). When the PLL enters this dead zone, the gain of the phase detector drops to zero, thereby breaking the loop and sending the PLL into a hunting or wandering behavior. Worse, since feedback has been broken, the PLL cannot shrink the noise of the VCO, thus behaving in a high jitter/high phase noise fashion.

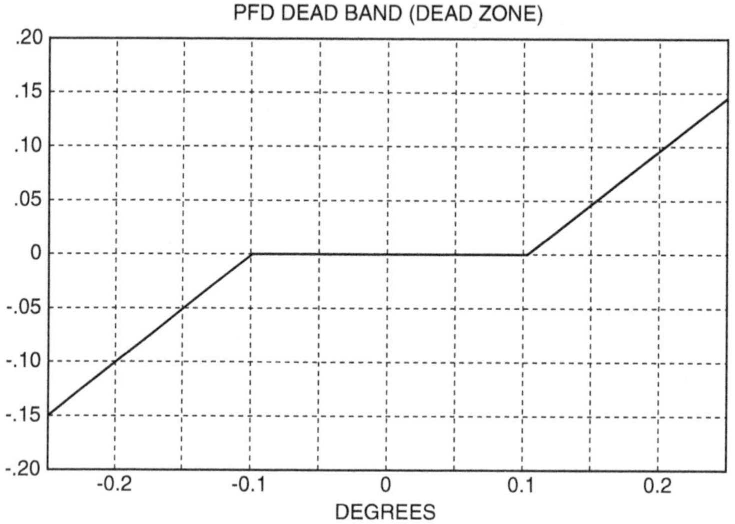

FIGURE 18.26 Type 4 PFD.

A typical dead zone phase angle range is shown in Figure 18.26 for the circuit of Figure 2.7 using the 74FXXX series logic family at 1 MHz operating frequency. No output from the phase detector occurs between -0.1 and $+0.1°$. This may seem small, but at 1 MHz operating frequency this is +/−278 ps (simulated using models

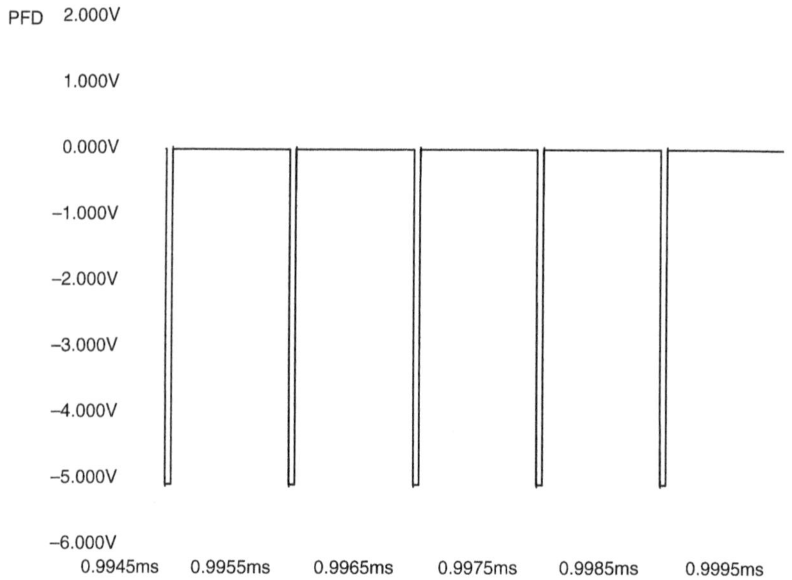

FIGURE 18.27 DC offset summed with PFD output causes reference sidebands' elevation.

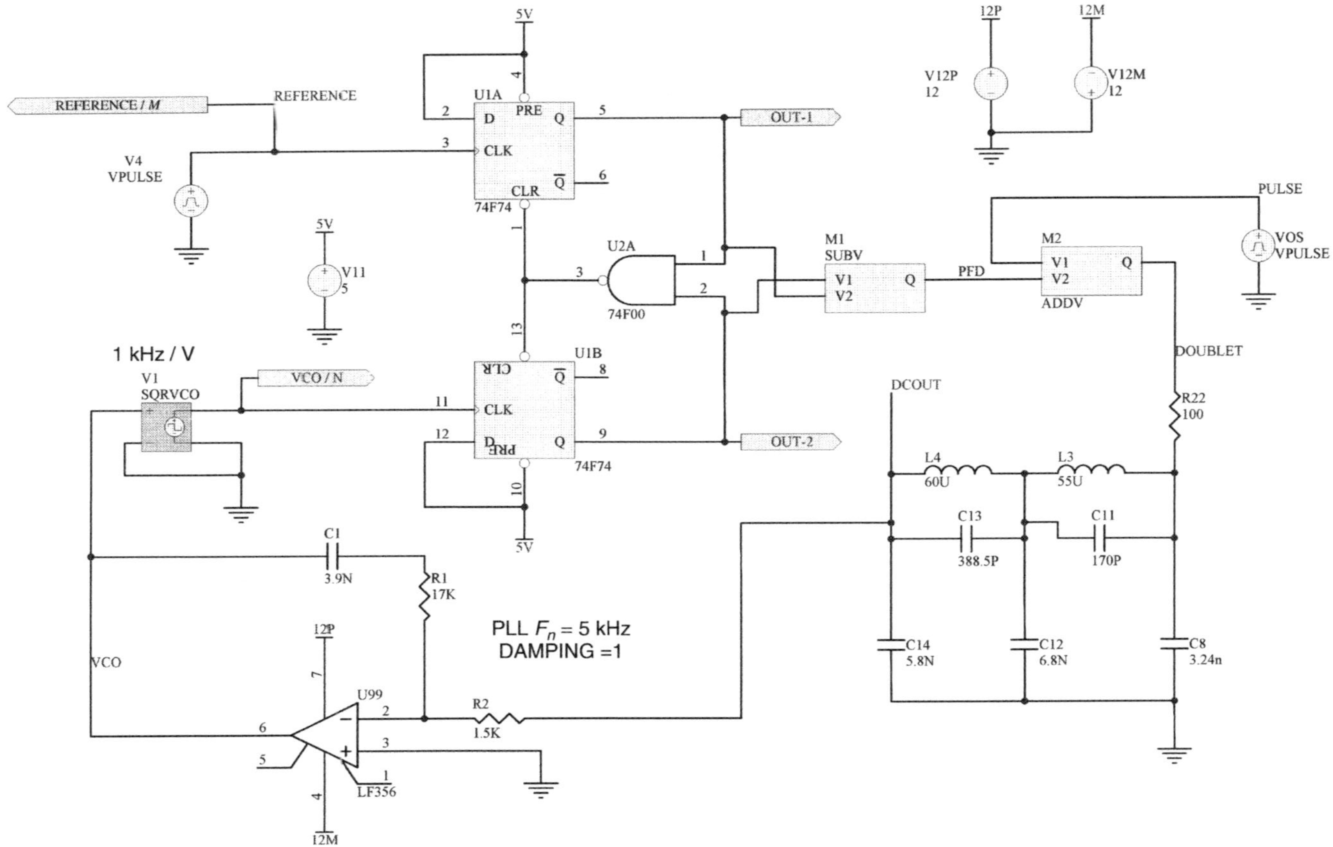

FIGURE 18.28 Summing a narrow pulse into the PFD output.

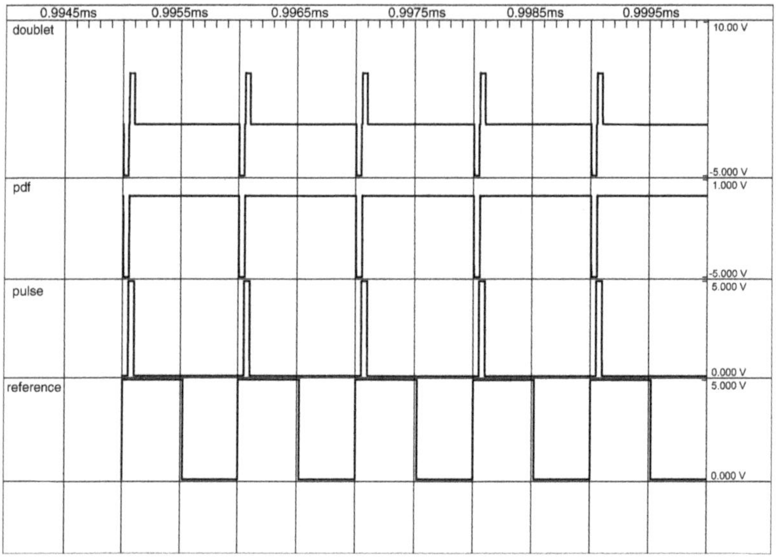

FIGURE 18.29 Pulse doublet resulting from narrow pulse summed into PFD output when PLL is operating closed loop.

of the 74 F logic components) of timing slip or hunting before the loop "catches up," which probably means even greater hunting than the 278 ps number. Just outside the dead zone, the curve is jagged and the PLL exhibits nervousness (not shown in the illustration). Well outside the dead zone, the phase detector becomes quite linear (slopes shown).

There have been a few cures for this problem. One cure is to simply insert a DC offset voltage in summation with the phase detector output, forcing the loop to negate said offset, thus forcing the phase angle into a region away from the dead zone. Such a brute force technique, however, also causes high reference sidebands. See Figure 18.27.

A better cure is to sum a narrow pulse into the phase detector output, triggered from the reference signal entering the phase detector. See Figure 18.28. The PLL then sees the DC average energy from this pulse and via feedback negates that DC, of course, by producing an equal and opposite pulse from the phase detector. This results in a pulse doublet, as shown in Figure 18.29 , which has zero average value and very nearly zero AC value. As a consequence, the resulting sidebands are not as significant in amplitude as they would have been with the pure DC offset technique. The pulse insertion technique was invented by Alfke et al., and assigned to Fairchild and patented many years ago [34]. There since have been many variations but in the same essential direction.

Figure 18.30 shows a simplified realization of the PFD of Figure 2.7. Note the use of readily available discrete logic, which could be replaced by custom incorporation of firmware or by higher speed logic, such as the ON Semiconductor® GigaCom silicon-germanium (SiGe) logic family or a custom equivalent or ASIC internal cell.

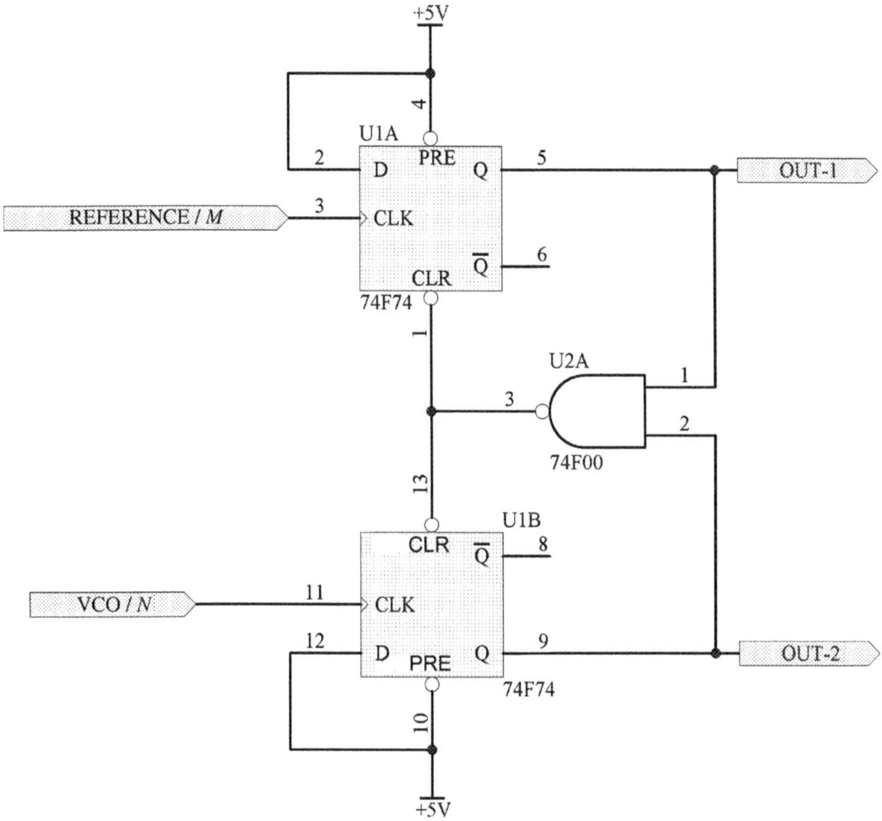

FIGURE 18.30 A simplified realization of a type 4 PFD.

18.11 SIDEBAND ENERGY DUE TO DC OFFSET FOLLOWING PHASE DETECTOR

How do we compute the reference sideband amplitudes when a DC offset is present superimposed on the PFD output? Here is a step-by step "back of the envelope" procedure to estimate the amplitude of the first (fundamental) pair of sidebands.

(1) Recall the damping coefficient (not the PLL damping factor) equal to the high-frequency shelf gain of the loop filter (R_f/R_{in}) and call it A1.

(2) Call the DC offset V_{os}. For small offsets, this will determine the peak value of the first sideband measured at the PFD output (because the PLL will produce a counteracting pulse of small duty factor to cancel this offset if the loop is type 2).

(3) Call the VCO gain in Hertz per volt $= K_{vco}$.

(4) Calculate the multiplied product of A1 and V_{os} and K_{vco} and this product equals the quantity $\Delta f = \{A1\}\{V_{os}\}\{K_{vco}\}$ = the peak FM deviation of the VCO (assuming the reference frequency is above the PLL passband).

(5) Apply Equation 8.2 to compute the peak radians of phase modulation. Recall Equation 8.2 states that $\beta = \Delta f / f_m$ where f_m is the reference frequency at the phase detector.

(6) Calculate this phase modulation in decibels as $20\,log_{10}\beta$ decibels relative to a radian, or dBr. Then each sideband is equal to 6 dB less than this number (recall from earlier discussion that each sideband is $\beta/2$). This final number is the amplitude of each of the pair of sidebands closest to and straddling the carrier as observed on a spectrum analyzer in dBc (decibels relative to carrier).

EXAMPLE

Suppose the op-amp following the PFD has a 20 mV DC offset *referred to the PFD output*. This forces the PFD to cancel this offset by producing a short pulse of the same average value. The peak voltage of the fundamental of this pulse is about 20 mV. Let us assume the loop integrator has a damping resistor of 3 k ohms and an input resistor of 1500 ohms, so that A1 = 2. Let us also assume a reference frequency of 10 MHz (well above a PLL passband of 200 kHz, for example) and a VCO of 1 MHz per volt tuning sensitivity (K_{vco}). Then, the peak radian phase modulation at a 10 MHz modulating rate will be $20\,log_{10}\beta = -48$ dBr or −54 dBc for each of the two sidebands located at +/−10 MHz around the carrier. Of course, there will be harmonics of the reference frequency as well, but these become rolled off at 6 dB/octave or more due to Equation 8.2, so we can expect the +/−20 MHz offset sidebands to be −60 dBc or lower. All of this presumes no sideband suppression filter. A Cauer filter having 40 dB suppression at 10 MHz and above will drop these sideband amplitudes by 40 dB while probably maintaining loop stability (a 200 kHz BW PLL, for example, would see trivial phase margin intrusion by a 4 or 5 MHz Cauer filter satisfying the 40 dB suppression at 10 MHz and above).

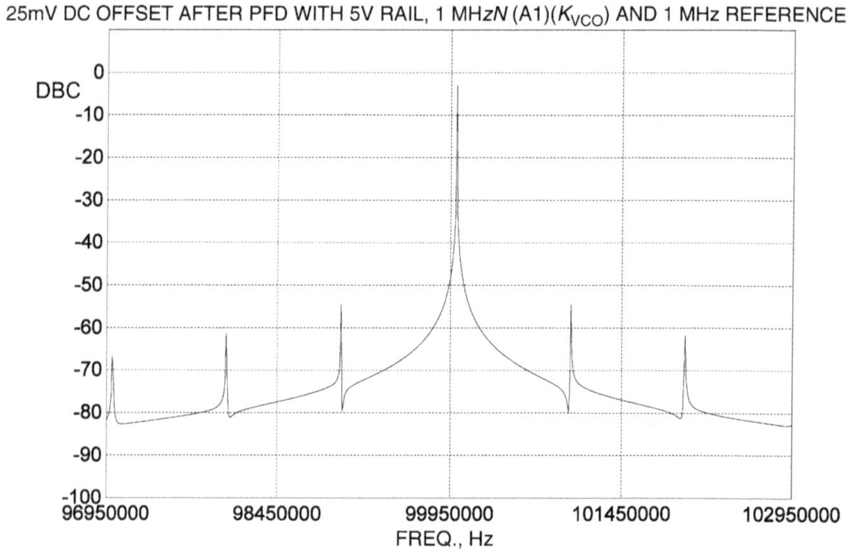

FIGURE 18.31 Example of offset post-PFD causing calculable reference sidebands.

An example of reference sidebands caused by a 25 mV DC offset following the PFD is shown in Figure 18.31 for a 100 MHz VCO having 100 kHz/volt K_{vco} and a loop integrator high-frequency shelf gain (A1) of 10 (for a product of 1 MHz/V) and a DC PFD rail of 5 V (VCC). The first sideband pair is exactly where predicted (−52 dBc) and higher order sidebands taper off at −6 dB per octave away from the carrier.

18.12 BRUTE FORCE PLL FREQUENCY ACQUISITION VIA SPEEDUP

The PLL itself can be temporarily modified to hasten frequency acquisition. For example, during power-up and/or frequency reprogramming, the loop component values may be momentarily altered to provide fast locking speed and then reset to their optimum values (either after dead reckoning or lock detection) to obtain the desired steady-state dynamics.

One technique is to make the loop integrator input resistor nonlinear by adding back-to-back diodes and a second input resistor. Another technique involves switching in extra components during a frequency reprogramming of the feedback divider [36]. Ideally, the value of the damping resistor should be nonlinear or switched in similar fashion. These techniques are well covered by Gardner [1] and in Refs [23,36]. In most of these techniques, the objective is to temporarily widen the PLL passband to permit higher gain to the beat waveform, or to increase the loop integrator speed in the applications where a PFD is used. When a PFD is not used, better acquisition performance is usually obtained with a quadricorrelator. Of course, we must qualify the term "better" to mean best compromise between short-term and long-term settling time and widest pull-in range. The "FastLock" technique employed by National Semiconductor® is popular in IC PLLs employing a PFD [37] (see toolkit file "FastLock_Technique∗.pdf"). Other firms have employed adaptations.

18.13 SHORT-TERM AND LONG-TERM SETTLING

The settling time behavior of a PLL is difficult to express unless we agree on whether our primary interest is in short-term or long-term settling. Obviously for very low damping factors, the PLL will reach its goal very quickly but then depart from it (overshoot and ring). A higher damping factor will take more time to reach the goal but will overshoot and ring less. Egan [7] shows that there is an optimum damping factor for best long-term settling, and that value is unity damping factor for a second-order type 2 loop [7].

18.14 *N*-OVER-*M* SYNTHESIS

There is a technique that can prove desirable for achieving low phase noise and settling time (using the highest reference frequency possible to avoid large feedback division ratios) called "*N*-over-*M*" synthesis. It merely approximates the desired

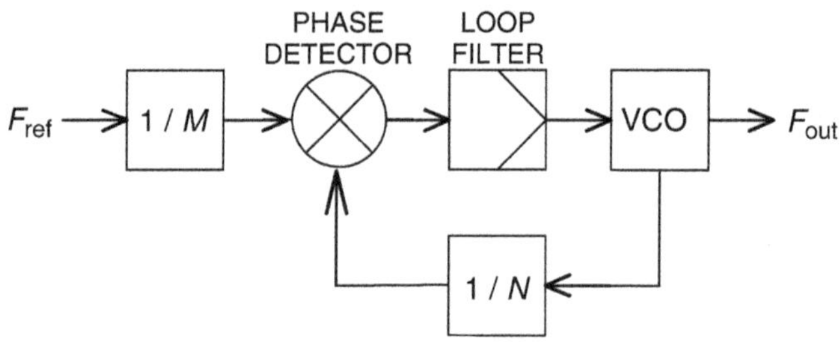

FIGURE 18.32 Block diagram of *N*-over-*M* frequency synthesizer.

output frequency rather than settling to a zero-error target. A small program is included in the toolkit called "n_over_m.exe" in the "Handy Tools" folder. Figure 18.32 shows a typical PLL frequency synthesizer block diagram. A reference signal is frequency divided by M and applied as an input to the PFD or phase detector. The other phase detector input is the VCO signal frequency divided by N. The output frequency $F_{\mathrm{out}} = N\{F_{\mathrm{ref}}/M\}$. If we want an integer PLL to have frequency steps as small as 1 MHz and have a 10 MHz reference, we must set $M = 10$. But if we are willing to just approximate each frequency, we can do a little better, that is, we can use smaller values for M, which has the advantages of (a) higher sideband frequencies that are more easily suppressed, (b) lower phase noise because N will be smaller, and (c) potentially faster settling (greater BW allowed).

Let's take an example. Suppose we have a 10 MHz crystal reference oscillator, and want to generate an output frequency of 1000.1 MHz. Use the free program in the toolkit, "n_over_m.exe", and find that for zero error, $N = 10001$ and $M = 100$ and the phase detector operating frequency is 100 kHz. Next, rerun the program and this time permit an error of 50 kHz. The new values are $M = 67$ and $N = 6701$ and a much higher phase detector operating frequency of 149.2537 kHz.

The technique works even better as we choose another scenario. Suppose, we want an output frequency of 1433 MHz. Then $M = 10$ and $N = 1433$ for zero error, but for an allowable error of 350 kHz, $M = 3$ and $N = 430$. That is an improvement in divider-caused phase noise of almost 10 dB and a phase detector (if a PFD is used) SNR degradation of 5 dB for a net win of 5 dB. Using all-analog techniques, the advantage would be the 10 dB number.

Applications where the error of *N*-over-*M* synthesis does not usually matter include radar, missiles, and unregulated communications. It may also be useful in consumer electronics and data reception as a local oscillator when paired with AFC (automatic frequency control) or APC (automatic phase control or derotating or despinning) downstream.

19

INJECTION PULLING OF MULTIPLE VCO's AS IN A SERDES

19.1 ALLOWABLE COUPLING BETWEEN ANY TWO VCOs VERSUS Q AND BW

In a previous chapter, we mentioned the work of Adler concerning injection locking. The phenomenon of injection pulling is also a concern, especially when it is enclosed by a PLL. The PLL locking angle and the injection pulling angle are not necessarily compatible, which results in a behavior resembling instability. The paper by Razavi [38] suggests that the incompatibility between PLL locking and injection pulling can be diminished by increasing the PLL bandwidth and the Q of the victim VCO.

Through simulation, we find that a relationship similar to that of Figure 19.1 exists. The simulation assumes the worst case phasing between the PLL and the injection pulling.

For one particular phase angle and its complement (depending on the type of phase detector), the PLL bandwidth will be abnormal. To avoid a variable "effective bandwidth," the cross-talk should be even lower than the number obtained from the graph, perhaps −60 dB or so.

19.2 TOPOLOGY SUGGESTION FOR ELIMINATING THE INJECTION PULLING

There is indeed a topology that highly immunizes a design from the cross-talk problem. If the VCO were tankless, the cross-talk coupling between sources would

Frequency Acquisition Techniques for Phase Locked Loops, Daniel B. Talbot.

Published 2012 by John Wiley & Sons, Inc.

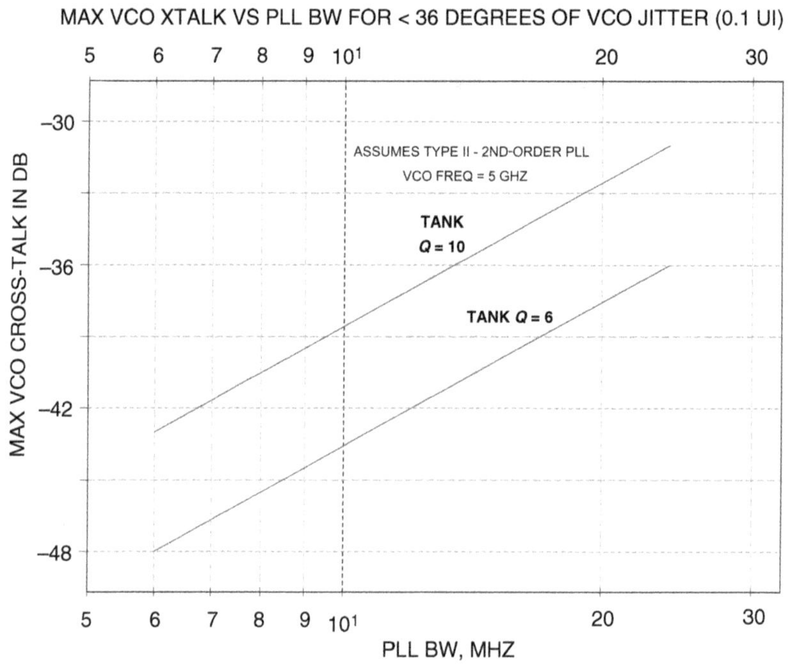

FIGURE 19.1 Graph of allowable cross-talk between VCOs in mutually leaking PLLs in order to minimize jitter to 0.1 UI versus BW in MHz.

be reduced from the nefarious enthusiasm of injection pulling to the benign simplicity of an algebraic spur.

A pseudo-VCO can be realized by mixing (heterodyning) a fixed frequency source from a single external tank (or better yet from a frequency-multiplied fixed crystal oscillator) with a low-frequency VCO, perhaps digitally generated using DDS techniques. The injection pulling phenomenon between DDS sources does not exist.

As an alternative, complex frequency heterodyning could be used, using a Cartesian phase modulator driven by a complex low-frequency source whose bandwidth merely need be wide enough to the PLL control signal without significant phase margin intrusion. An example of a Cartesian phase modulator is shown in Figure 19.2.

The "external carrier" might be a 5 GHZ, for example, signal distributed to all SerDes channels. Then, the remaining circuitry of Figure 19.2 would be on-chip, replicated for each channel. The DDS output circuitry runs at a very low frequency, say 0–750 kHz, and the output of the DDS is complex, so that the signs of either real or imaginary terms can be inverted (effectively heterodyning by either positive or negative offset frequency). Thus, each DDS can be clocked much higher, and the real and imaginary output image suppression low-pass filters can be matched with active filters on-chip (not shown). The result is a "VCO" having, for example, a 5 GHz center frequency and +/−750 kHz variability around that center. Although it

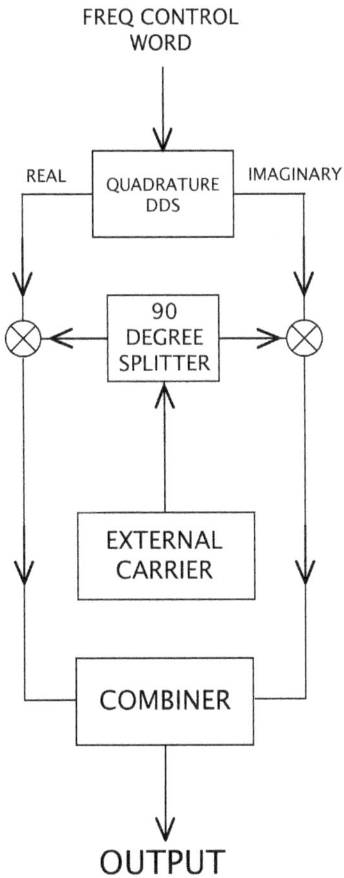

FIGURE 19.2 Cartesian coordinate phase modulator makes virtual VCO without using tank circuits (which are vulnerable to injection pulling) for a multiple oscillator application, such as a SerDes.

seems more complex than the high-frequency LC VCO, remember that an on-chip inductor is a large component, and the proposed circuitry is robustly immune to injection locking or pulling. This removes the injection risk entirely in a multiple oscillator SerDes. It is not known whether this technique has been implemented commercially.

20

DIGITAL PLL EXAMPLE

There are occasions when it would be insane to attempt an analog approach to a PLL design, such as when the goal is a 0.1 Hz bandwidth. Run the file "pllsynth.exe" in the toolkit, previously discussed. Assume a phase detector having 1 V per radian sensitivity, and a VCXO (voltage-controlled crystal oscillator) having 1 kHz/volt tuning sensitivity and a factor 10 feedback divider, and a 0.1 Hz natural frequency with unity damping. Then from the program, the loop integrator capacitor must be >1 farad and for a 1 mHz bandwidth, the capacitor size skyrockets. Such small PLL bandwidths are typically employed for data clock repeaters where jitter buildup accumulates after many repeaters. A more compact and physically realizable approach uses a digital accumulator in place of the analog integrator, with a detour path for damping (digital multiplier to set the gain) and a DAC (digital-to-analog converter) to drive the VCXO (or skip the DAC and drive a DSP–DDS simulated VCO).

See Figure 20.1 for a closed loop frequency response of a 1 mHz PLL, whose block diagram is shown in Figure 20.2, and whose SPICE model is shown in Figure 20.3.

Below is the *basic* computer program used to compute the weights of coefficients for the accumulator (loop integrator) and damping multiplier. The program calculates the absurd component values for an analog integrator as well as the coefficients for a digital implementation. The damping multiplier and accumulator can be implemented in a DSP (digital signal processing) chip or perhaps in an FPGA (field programmable gate array).

Frequency Acquisition Techniques for Phase Locked Loops, Daniel B. Talbot.

Published 2012 by John Wiley & Sons, Inc.

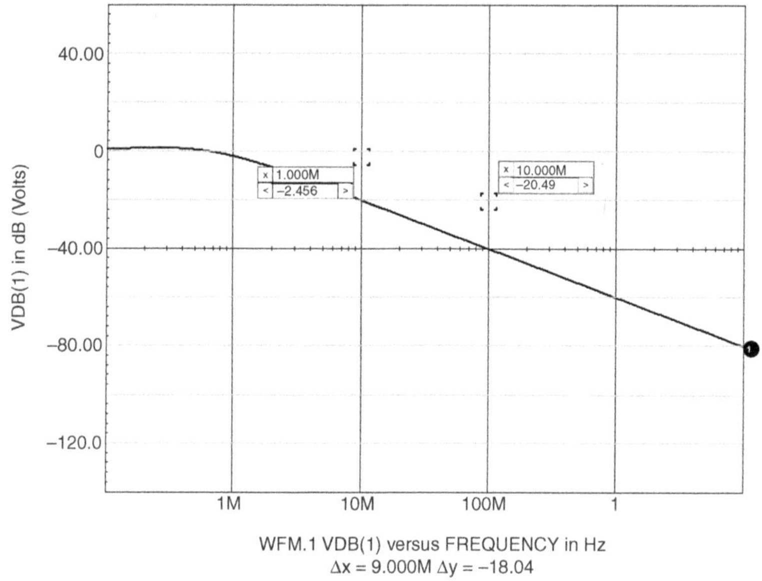

FIGURE 20.1 Closed loop frequency response for a 1 mHz PLL.

The phase detector output rails are expressed in two places: the detector itself as having a peak-to-peak radian excursion, and as A/D (analog to digital) corresponding fully loaded bit width. The other inputs to the program are self-explanatory from the block diagram.

A PLL with only 100 mHz bandwidth (natural frequency) is actually used in sophisticated telecom repeaters. A very low drift-rate VCO must be used or else the PLL will be constantly chasing the VCO drift and failing to remain in lock.

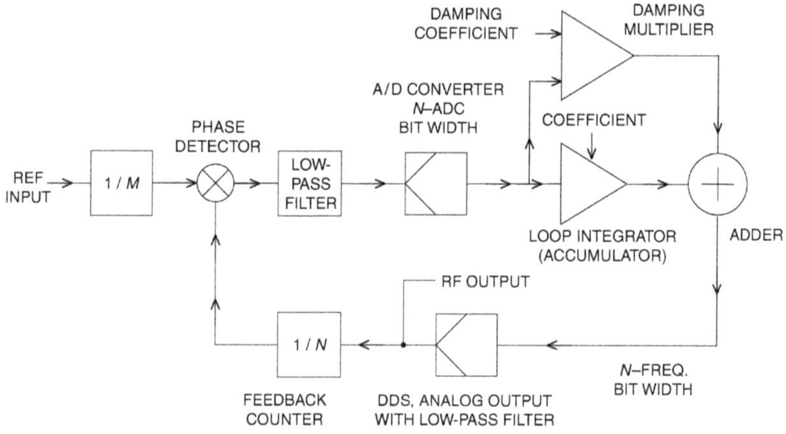

FIGURE 20.2 Block diagram of the digital PLL.

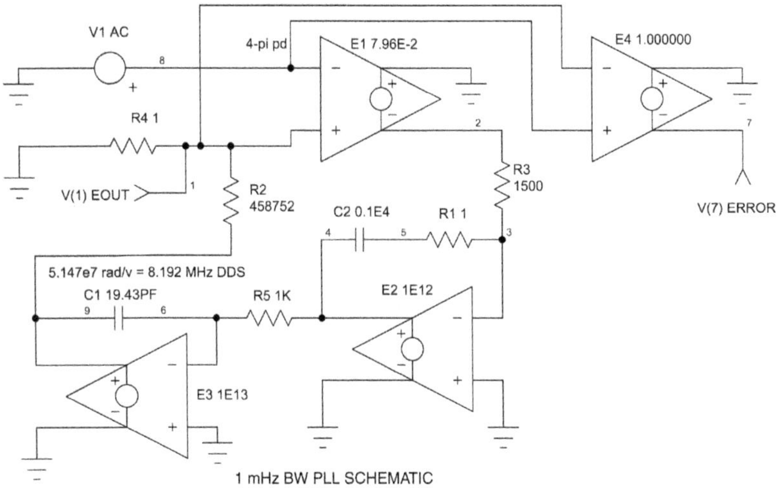

FIGURE 20.3 Spice model for a 1 mHz PLL.

```
CLS
PRINT: PRINT
INPUT "vco dds max freq, full scale, hz ="; v
INPUT "dds freq code width =", nfreq
INPUT "phase detector a/d output code width =", nadc
INPUT "phase detector PK-PK radians ="; p1
p1 = 1/p1
v = v * p1
INPUT "required damping factor d ="; d
INPUT "required loop natural frequency f0 ="; f0
INPUT "Frequency Division ratio n = "; x1
INPUT "numerical integrator time step (per accumulator, sec =";
tstep
PRINT "accumulator instruction rate therefore = "; 1/tstep
PRINT "for the case where the division ratio N =";
PRINT x1
twopi = 2 * 3.1459
cf = v/ (twopi * f0 * f0 * x1 * 1500!)
rf = 2 * d/ (twopi * f0 * cf)
v = v/p1
PRINT " cf = ";: PRINT cf;: PRINT " farads "
PRINT " rin= ";: PRINT " 1.5 k ohms "
PRINT " rf = ";: PRINT rf;: PRINT " ohhms"
damp = rf/1500
accw = 2 ^ nfreq * tstep/ (1500 * cf * 2 ^ nadc)
PRINT " accumulator input weight coefficient ="; accw
```

```
PRINT "damping weighting coeffficient ="; damp * (2 ^ nfreq/2 ^ nadc)
PRINT "pull-in range for no cycle-slipping is approx =";: PRINT x1 *
f0;:
PRINT "hz"
END: STOP: END
```

The step response settling time to a small phase step is thousands of seconds and to a large step is considerable. Therefore, it is presumed that a loop coefficient manipulation should be performed during start-up, akin to the analog speed-up techniques discussed earlier, or a digital implementation of a quadricorrelator with "disconnect."

21

CONCLUSION

It is hoped that this book has helped the reader understand the basics of frequency acquisition and its application. We have digressed into PLL design issues as a necessary part of this process to insure that the reader will know both sides of the interface: the acquisition side of the circuitry and the PLL side. There are many great works identified in the references as well as others that can be had with more search effort. We have avoided discussing detailed circuitry in most cases, since it is not known whether the design is intended to be implemented with GaAs, SiGe, or CMOS (or even discrete) devices. The reader will undoubtedly know how to implement the block diagrams we have presented, as well as comprehend our discussions on interferers and the approach to surviving them in the process of frequency acquisition. We have shown how simple DC offset can be effective in avoiding false locking on sidebands (data or other). We have shown how limiting can be used to insure better immunity from interferers if a quadricorrelator or frequency discriminator is used. We have shown that an interferer causes a nonsinusoidal beat waveform in a frequency discriminator or quadricorrelator. We have also cautioned that nonlinear circuitry should be avoided following the frequency discrimination, since asymmetry will distort the beat waveform and generate an unwanted DC offset, thereby causing convergence to the wrong frequency offset and possibly failure to handoff activity to the PLL so that it may finish the locking process.

We have described some basic PLL techniques, including N-over-M, integer, DDS-augmented, multiple loop, and frac-N synthesis, and we have included files in the toolkit that it is hoped become useful to the reader. We have chosen to maximize the use of free software (and identified its source) to allow the reader to

Frequency Acquisition Techniques for Phase Locked Loops, Daniel B. Talbot.

Published 2012 by John Wiley & Sons, Inc.

simulate most of the illustrations. Best wishes for your success, dear reader, in all of your PLL designs, and we hope we have been of some help, be it ever so simplistic. You are going to expand on these ideas to fit your needs, and hopefully enlighten the rest of us.

Many topics were given only short mention here, such as noise shaping in fractional-*N* PLLs, simply because such topics are covered in detail elsewhere.

Thank you for purchasing and reading this book. It is sincerely hoped that our discussions were clearly articulated and easily understood.

REFERENCES

[1] Gardner, Floyd M., *Phaselock Techniques*, Wiley, 1966, and subsequent editions, Library of Congress number 66-22837.

[2] Rohde, Ulrich L., *Digital PLL Frequency Synthesizers*, Prentice-Hall, 1983, Library of Congress number 82-11216.

[3] D'Azzo, John J. and Houpis, Constantine H., *Feedback Control Systems Analysis and Synthesis*, McGraw-Hill, 1966, Library of Congress number 65-17391.

[4] Driscoll, M.M., Phase noise performance of analog frequency dividers, *IEEE Trans. Ultrason. Ferroelectr. Freq. Control* 1990, 37 (4), 295–301.

[5] Millman, J. and Taub, H., *Pulse, Digital, and Switching Waveforms*, McGraw-Hill, 1965, Library of Congress number 64-66293.

[6] Adler, R., A study of locking phenomena in oscillators, *Proc. IRE* 1946, 351–357, reprinted in *Proc. IEEE*, October 1973, 61 (10), 1380–1385.

[7] Egan, William F., *Frequency Synthesis by Phase Lock*, Wiley Interscience, 1981.

[8] Skyworks Solutions, Inc., *Theory and Application of Sampling Phase Detector*, Application Note (included in the toolkit for this book).

[9] Park, C., Seo, H. and Kim, B., A Noise Optimized Passive Mixer for Charge-Domain Sampling Applications, *J. Electromagn. Waves Appl.* 2009, 23, 1909–1917 (included in the toolkit for this book).

[10] Agoston, *Tracking Sample and Hold Phase Detector*, US Patent 4751468, June 14, 1988, assigned to Tektronix.

[11] Klapper, J. and Frankle, J., *Phase-Locked and Frequency-Feedback Systems*, Academic Press, 1972, Library of Congress number 72-76546(out of print).

Frequency Acquisition Techniques for Phase Locked Loops, Daniel B. Talbot.

Published 2012 by John Wiley & Sons, Inc.

[12] Robins, W. P., *Phase Noise in Signal Sources (IEE Telecommunications Series 9)*, Institution of Engineering and Technology, 1984.

[13] Schwartz, M., *Information Transmission, Modulation, and Noise*, McGraw-Hill, 1959, Library of Congress number 59-9993.

[14] van derPol, B., On relaxation-oscillations, *Lond., Edinb. Dublin Phil. Mag. J. Sci.* 1927, 2(7), 978–992.

[15] Leeson, D.B., A simple model of feedback oscillator noise spectrum, *Proc. IEEE* 1966, 329–330.

[16] Baghdady, E. J., Theory of stronger-signal capture in FM reception, *Proc. IRE* 1958, 46, 728–738.

[17] Talbot, D. B., Improving TV Audio at the Broadcast Transmitter–Receiver Interface, *J. Audio Eng. Soc.* 27, 482–491 (1979 June), vol 26, no. 6.

[18] Crosby, M. G., Frequency modulation noise characteristics, *Proc. IRE* 1937, 25.

[19] Peregrine Semiconductor Corp., Normalized Phase Noise in UltraCMOS Devices, Staff written application note.

[20] Costas, John P., Synchronous Communications, *Proc. IEEE* 1956, republished in the same journal in vol. 90, no. 8, 1461–1466. August 2002 as classic paper.

[21] Holmes, Jack K., *Coherent Spread Spectrum Systems*, Wiley, 1982.

[22] Hedin, G., Holmes, J. K., Lindsey, W.C., and Woo, K.T., Theory of False Lock in Costas Loops, *IEEE Trans. Commun.* 1978, 26, 1–12.

[23] Kroupa, Venceslav F., *Phase Lock Loops and Frequency Synthesis*, Wiley, 2011.

[24] Day, J. R., Serrasoid F-M Modulator, *Electronics* 1948, 72–76.

[25] Scott, P. R., Large amplitude operation of the nonlinear oscillator, *Proc. IEEE* 1968, 56(12), 2182–2183.

[26] Narayanan, S., Transistor distortion using Volterra series representation, *Bell Syst. Tech. J.* 1967, 46, 991–1023.

[27] Volterra, V., *Theory of Functionals and of Integral and Integro-Differential Equations*, Dover, New York, 1959.

[28] Sagers, R.C. (Motorola), Intercept point and undesired responses, *IEEE Vehicular Technology Conference paper*, vol. 32, May1982, pp. 219–230.

[29] Setty, R, Ji, D., and Kaylie, H. (Mini-Circuits), *Figure of Merit of Mixer Intermod Performance*, Application Note AN-00-001, Mini-Circuits, Brooklyn, NY.

[30] Talbot, D., *A Review of Non-Linear Distortion Fundamentals*, paper presented at 76th convention of the Audio Engineering Society (AES), 2134(B-5) October 1984, New York, NY.

[31] Wadley, Trevor, (search Google for www.saiee.org.za/content.php?pageID=265#).

[32] Analog Devices Inc., *A Technical Tutorial on Digital Signal Synthesis*, 1999, paper 450968421.

[33] Goldberg, Bar-Giora, *Digital Frequency Synthesis Demystified*, LLH Technology Publishing, 1999.

[34] Alfke, Peter H., et al., *Phase-Locked Loop Frequency Synthesizer*, US Patent 4,023,116, May 10, 1977, assigned to Fairchild Camera and Instrument Corp.

[35] Bullock, Scott R., Control Theory Analyzes Phase-Locked Loops, *Microwaves RF Mag.* May 1992, 157–165.

[36] Roh, G.-T. et al., Optimum phase-acquisition technique for charge-pump PLL, *IEEE Trans. Circuits Sys.* 2002, 44(9), 729–739.

[37] Byrd, David, et al., *A Fast Locking Scheme for PLL Frequency Synthesizers*, National Semiconductor Application note 1000, July, 1995.

[38] Razavi, B., 2006, Mutual injection pulling between oscillators,IEEE 2006 Custom Integrated Circuits Conference (CICC), pp. 675–678.

INDEX

Frequency Acquisition Techniques for Phase Locked Loops, Daniel B. Talbot.

Published 2012 by John Wiley & Sons, Inc.

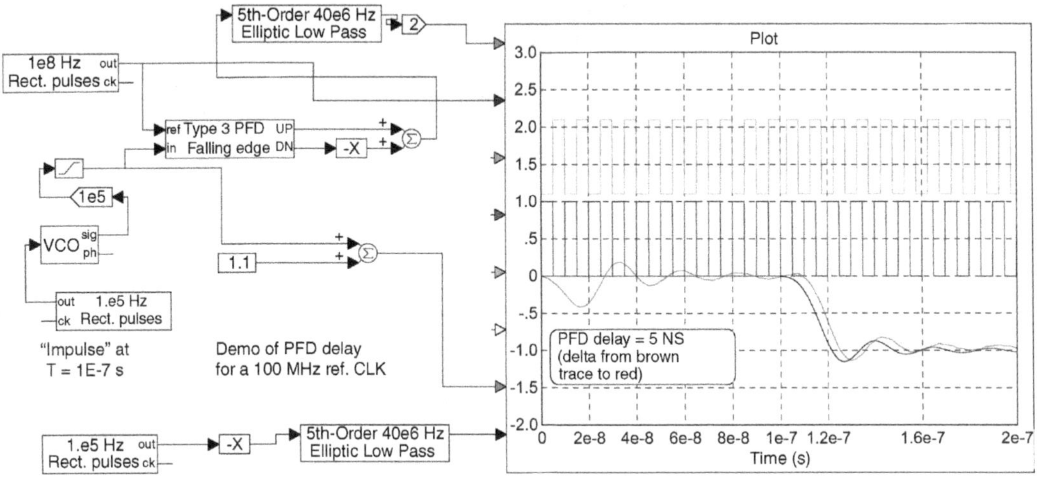

FIGURE 5.9 Phase detector delay (5 ns for a 10 ns interpulse period) = ½ period (most sampled-data phase detector exhibit similar delay).

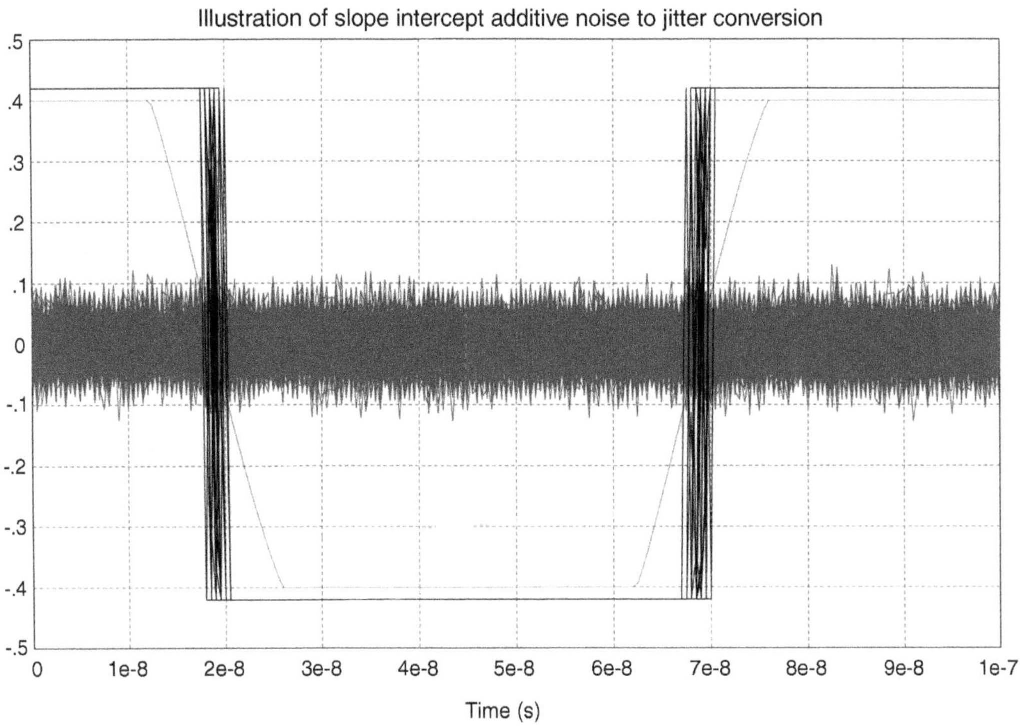

Red trace = Pulse applied to logic gate Blue trace = Noisy threshold of logic gate

Black trace = Final output from logic gate

FIGURE 9.4 Illustration of additive white Gaussian noise (AWGN) causing jitter via slope intercept.

Frequency Acquisition Techniques for Phase Locked Loops, Daniel Talbot.

Published 2012 by John Wiley & Sons, Inc.

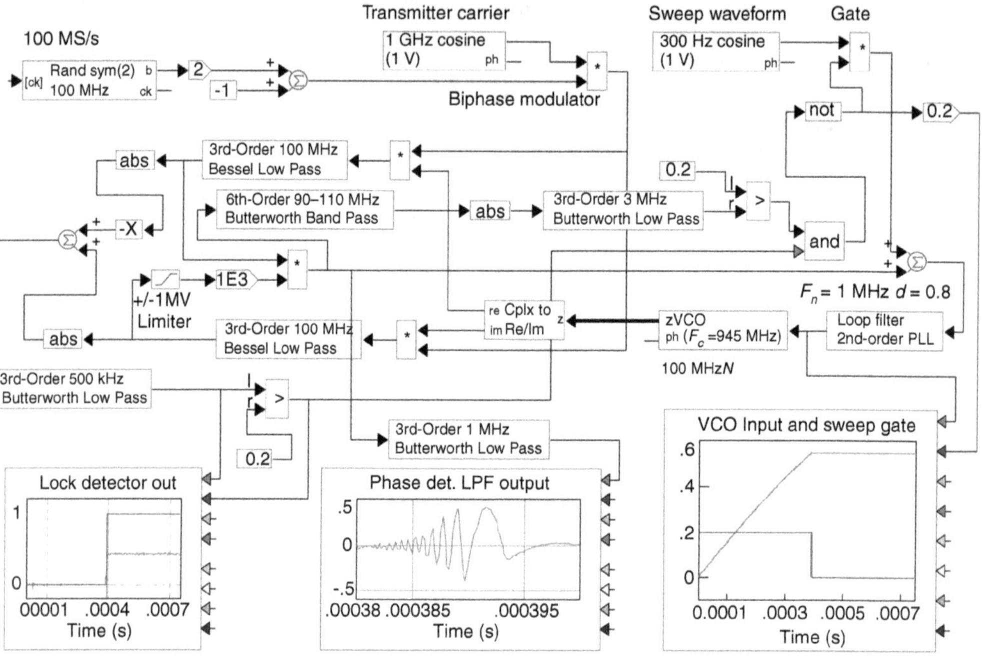

FIGURE 11.13 VisSim COMM 8 simulation of BPSK costas PLL with sweep, NRZ data and false lock prevention, coherent lock Detection.

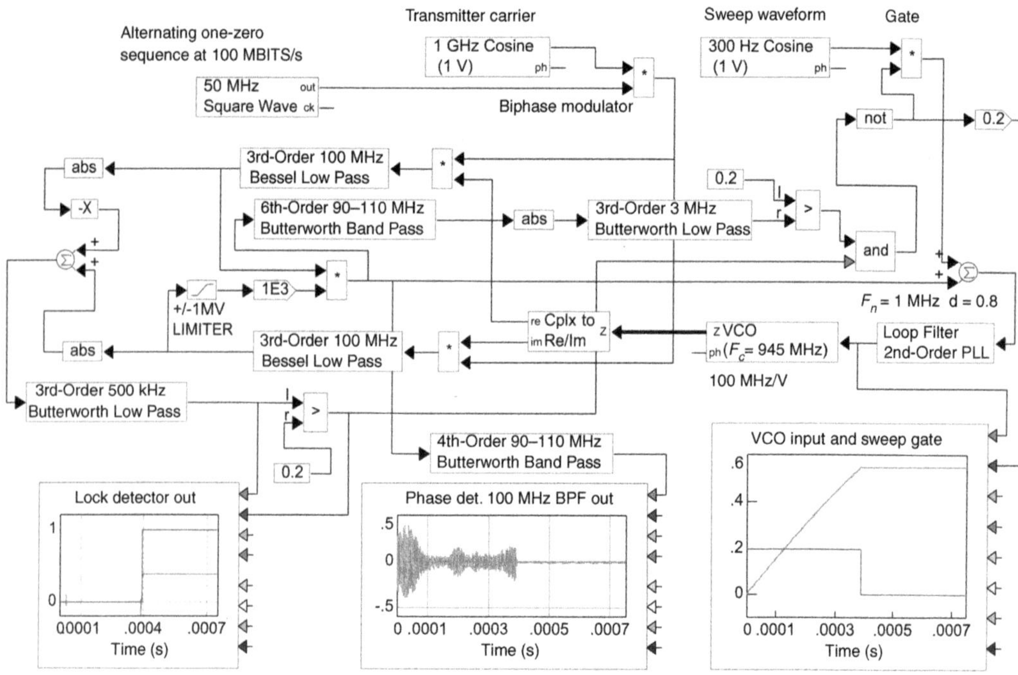

FIGURE 11.14 BPSK Costas PLL with false lock prevention responding to a 50 MHz square wave (equivalent TO 100 Mbits/sec) modulation (note strong 100 MHz from phase detector but not from lock detector as VCO sweep crosses 50 MHz offset Frequency).

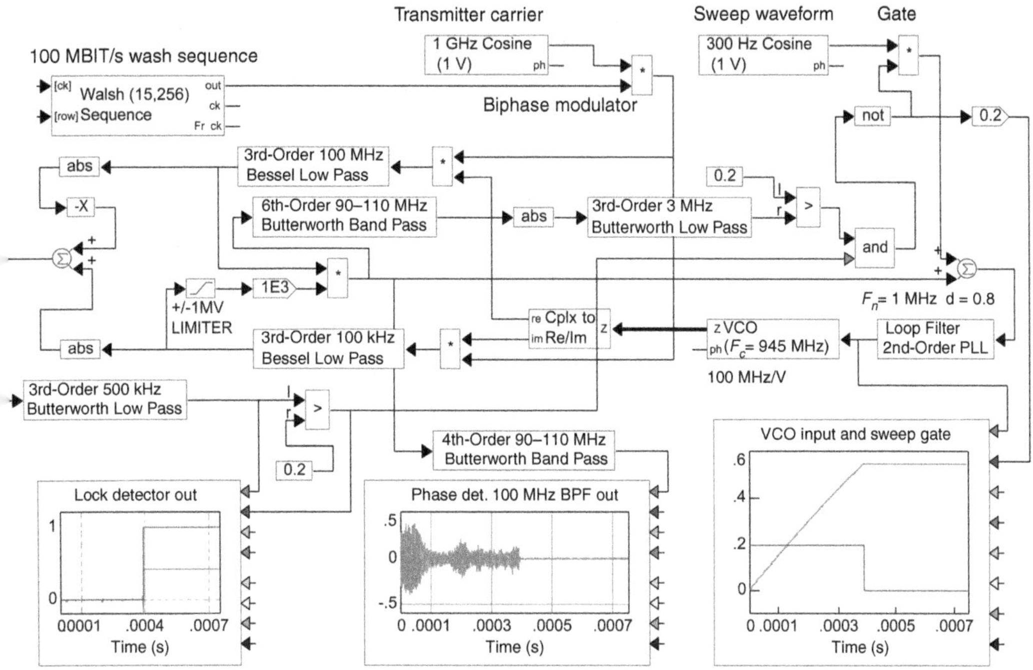

FIGURE 11.15 BPSK Costas PLL with false lock prevention responding to a walsh sequence (100 Mbits/s) modulation (note strong 100 MHz from phase detector but not from lock detector as sweep crosses 50 MHz offset frequency).

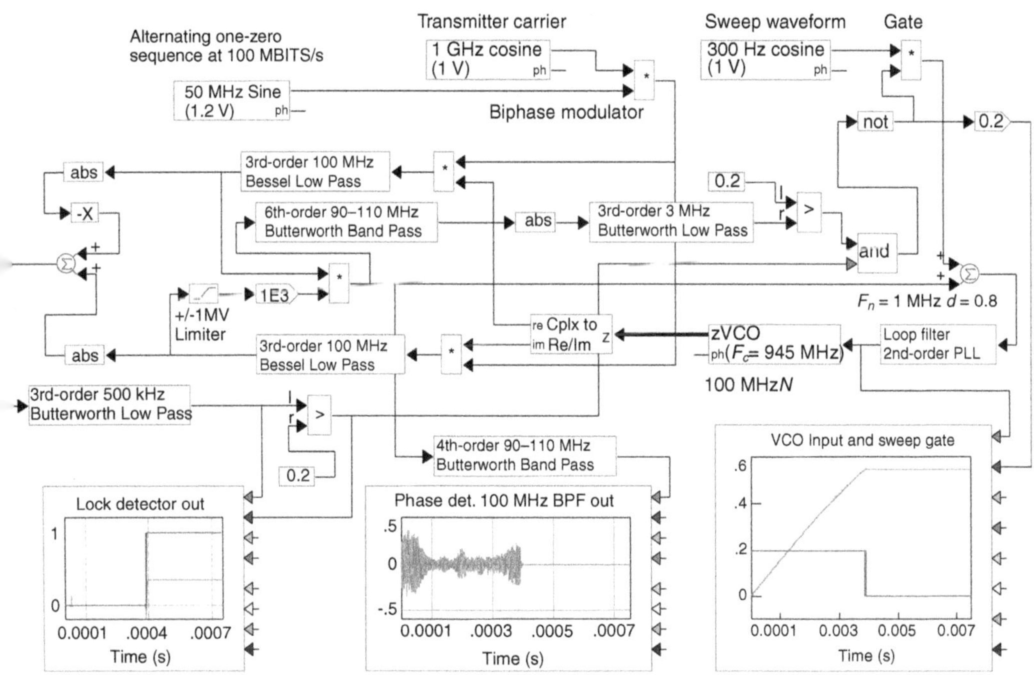

FIGURE 11.16 BPSK Costas PLL with false lock prevention responding to a 50 MHz (100 Mbits/s) sine wave modulation (note strong 100 MHz from phase detector but not from lock detector a sweep crosses 50 MHz offset frequency).

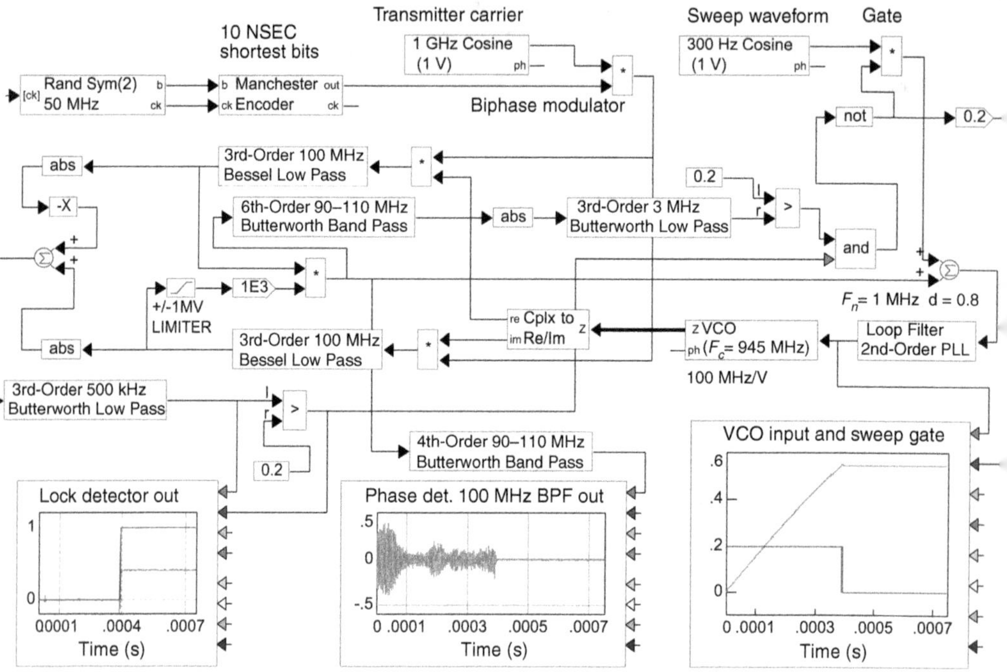

FIGURE 11.17 BPSK Costas PLL with false lock prevention responding to a 100 Mbits/s Manchester modulation (note strong 100 MHz from phase detector as sweep crosses 50 MHz offset frequency).

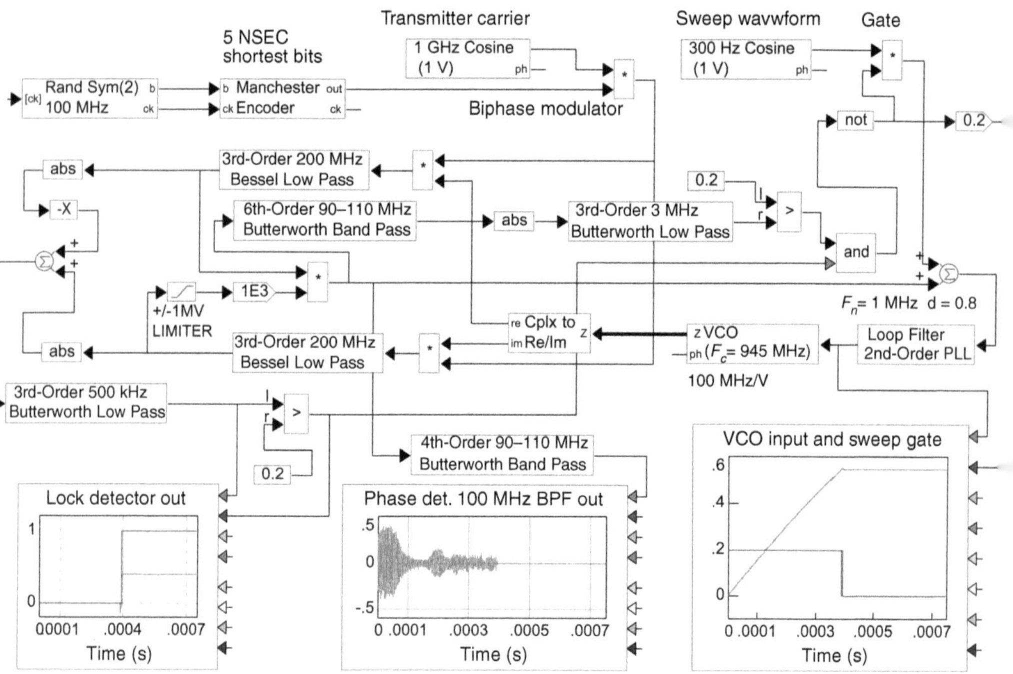

FIGURE 11.18 BPSK Costas PLL with false lock prevention responding to a 200 Mbits/s Manchester modulation (note strong 100 MHz from phase detector but not from lock detector as sweep crosses 50 MHz offset frequency).

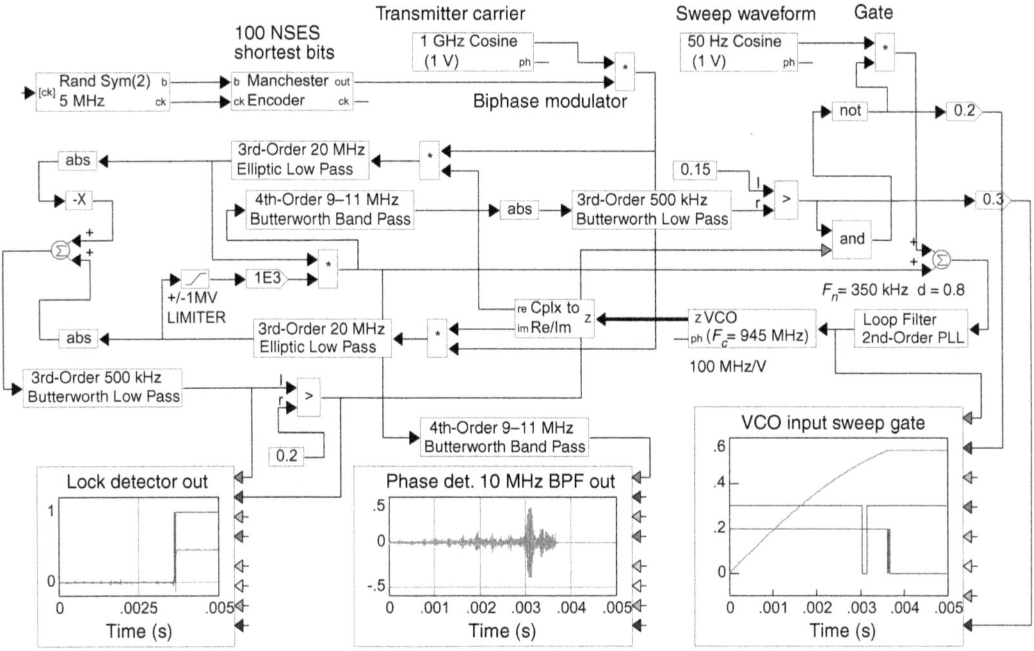

FIGURE 11.19 BPSK Costas PLL with false lock prevention and 10 Mbit/s Manchester modulation (note strong 10 MHz from phase detector but not from lock detector as sweep crosses 5 MHz offset frequency).

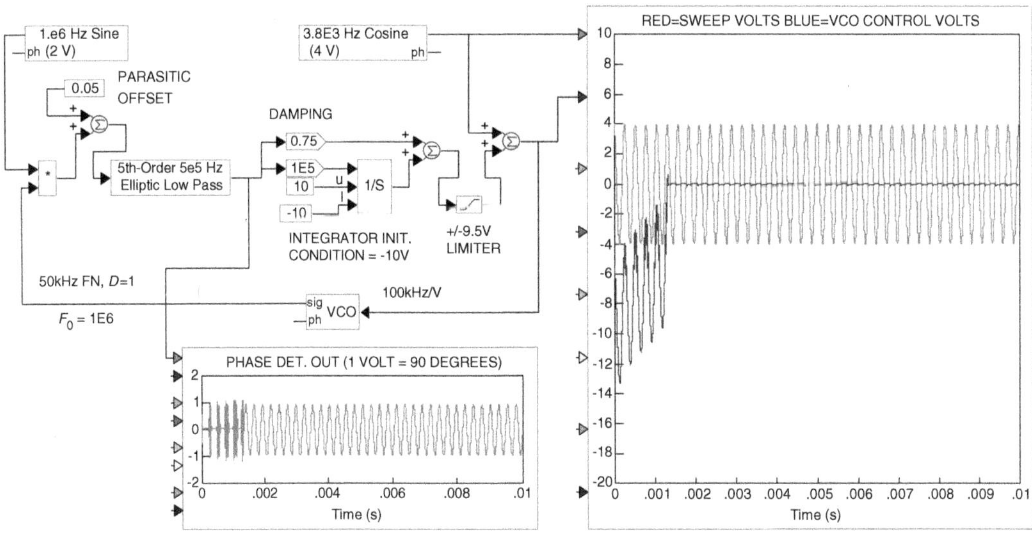

FIGURE 12.1 Maximum accelerating VCO sweep (note phase detector output shows peaks equal to 90° peak phase excursions).

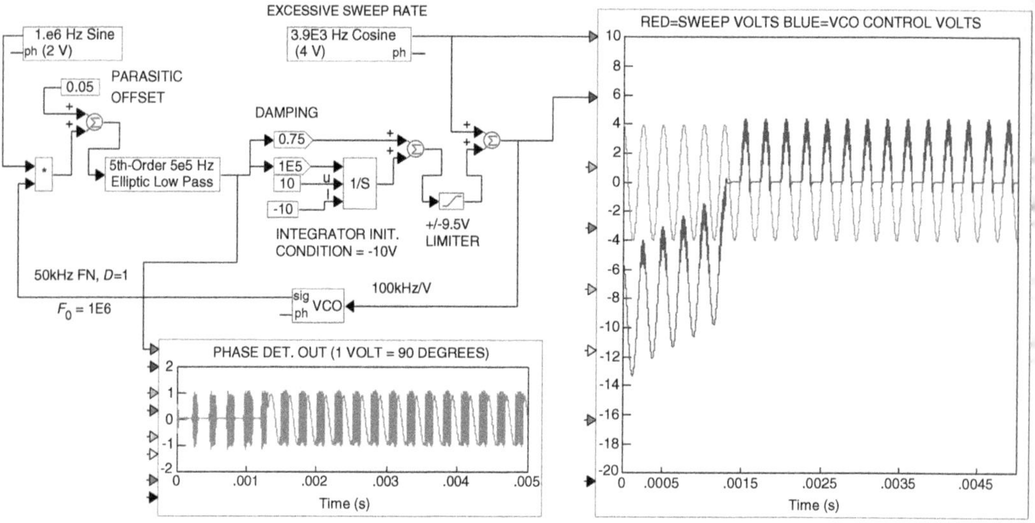

FIGURE 12.2 Excessive sweep rate (acceleration) causes phase detector rollovers (cycle slips) and splatter (positive peaks are expanded and negative peaks clipped due to phase detector's output voltage offset of 50 mV).

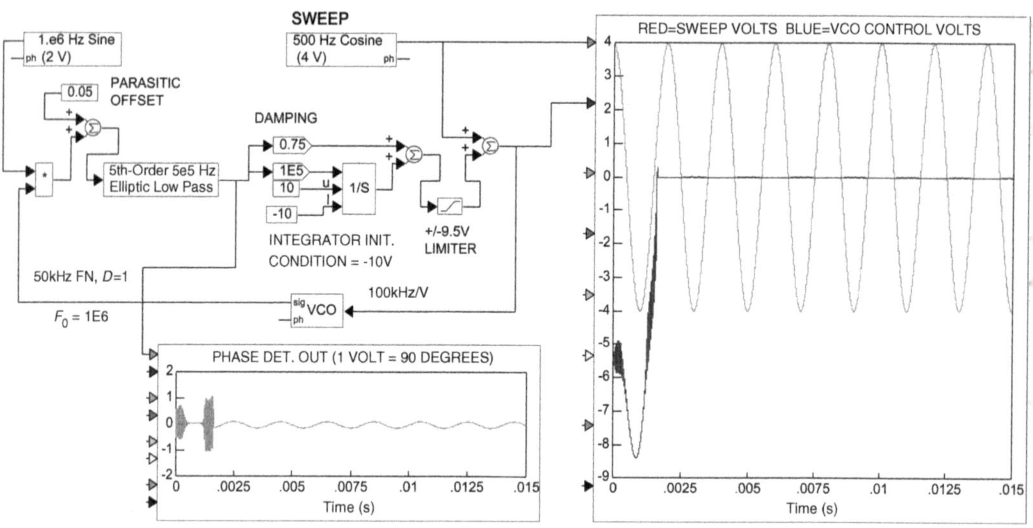

FIGURE 12.3 VCO sweep runs continuously; causes appreciable phase wobble of ~30° peak; also notice phase detector output offset to force integrator to drift in favorable direction (initially pegged at −10 V rail, damping pegged at −9.5 V); sweep rate = 500 Hz at 4 V peak = 1.257 GHz/s at sweep zero crossings.

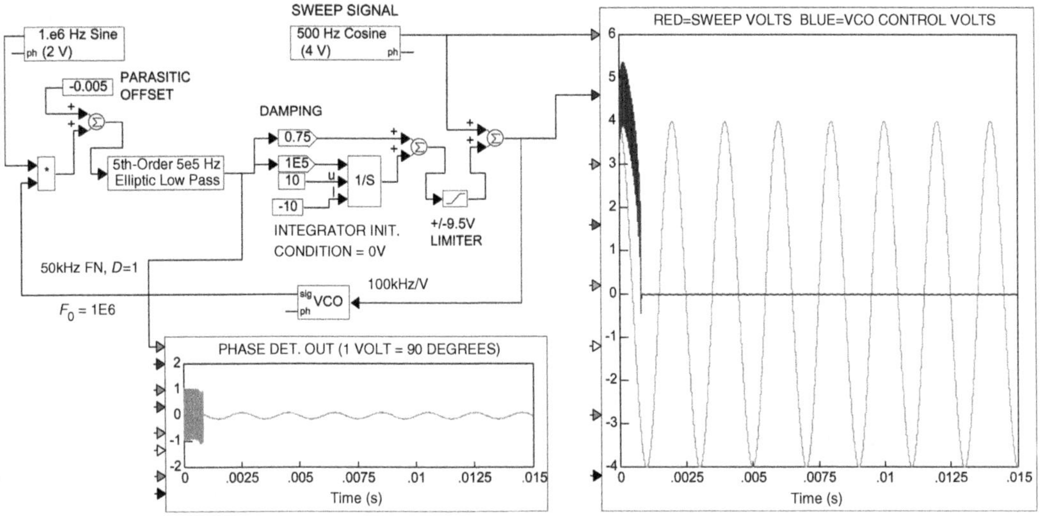

FIGURE 12.4 Integrator zero volts initial condition with −5 MV DC input offset, sweep signal continuously applied (no lock detector circuitry; note phase wobble of ~30° peak).

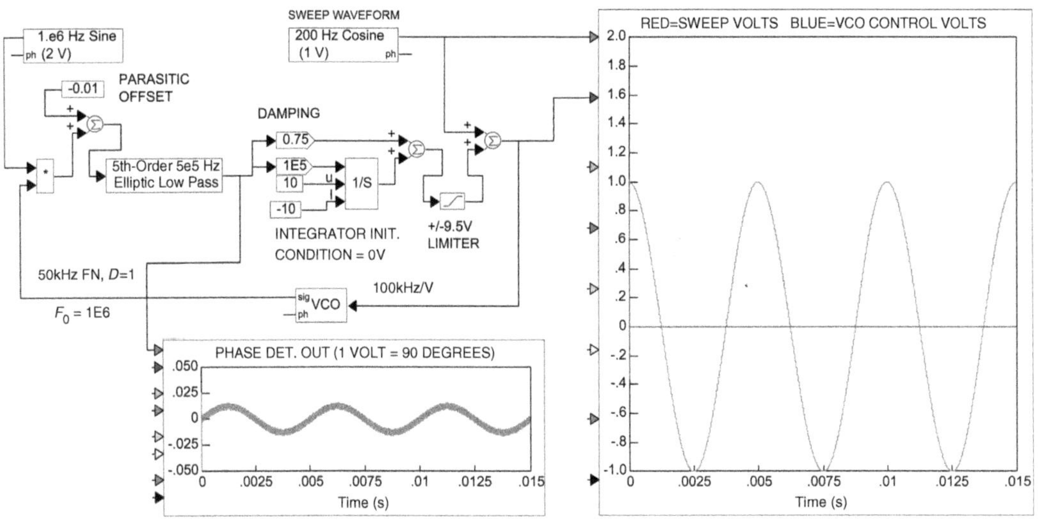

FIGURE 12.5 Comfortably slow sweep rate resulting in ~10° phase wobble when sweep runs continuously and not disconnected by a lock detection circuit (lucky acquisition, not recommended to combine type-II PLL with continuously running VCO sweeping).

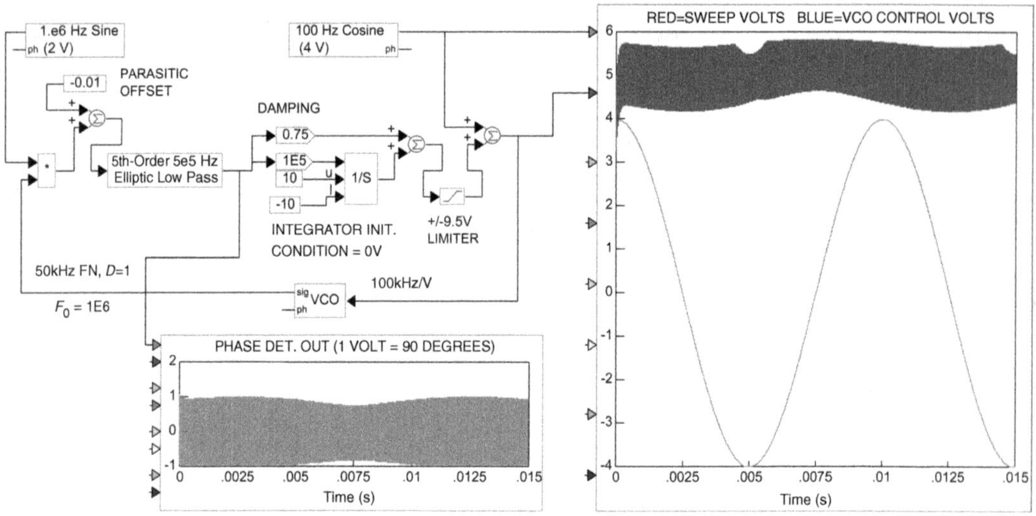

FIGURE 12.6 Unsuccessful phase lock in a type-II PLL with continuously running sweep due to false self-locking.

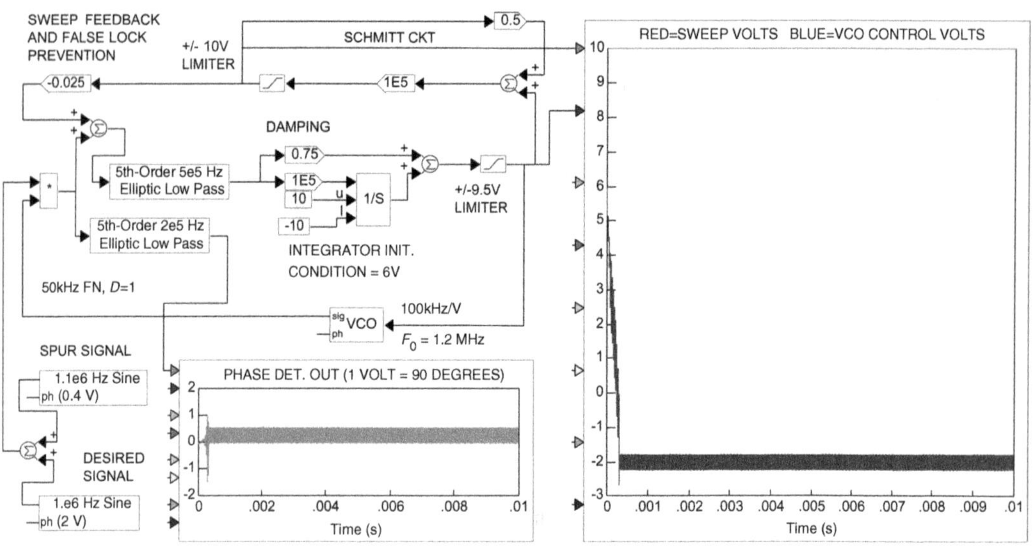

FIGURE 12.8 Automatic self-sweeping PLL having false lock prevention.

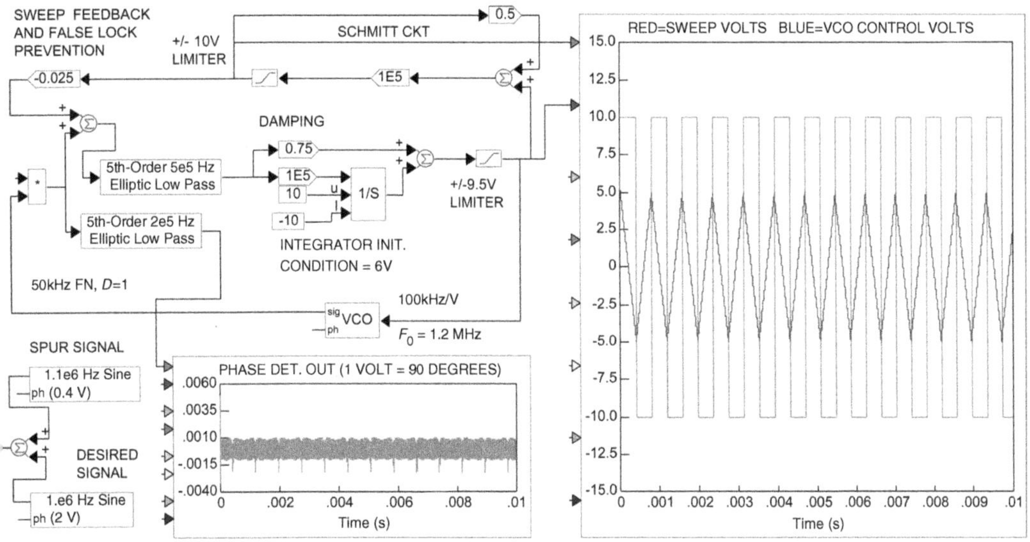

FIGURE 12.9 Automatic self-sweeping PLL having false lock prevention in no-signal sweeping mode.

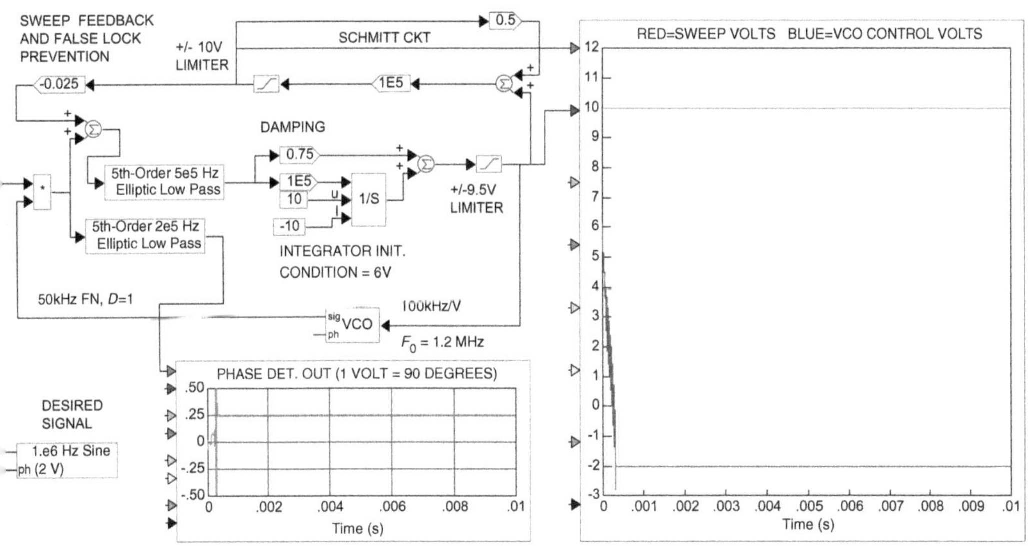

FIGURE 12.10 Automatic self-sweeping PLL having false lock prevention: no-spur behavior.

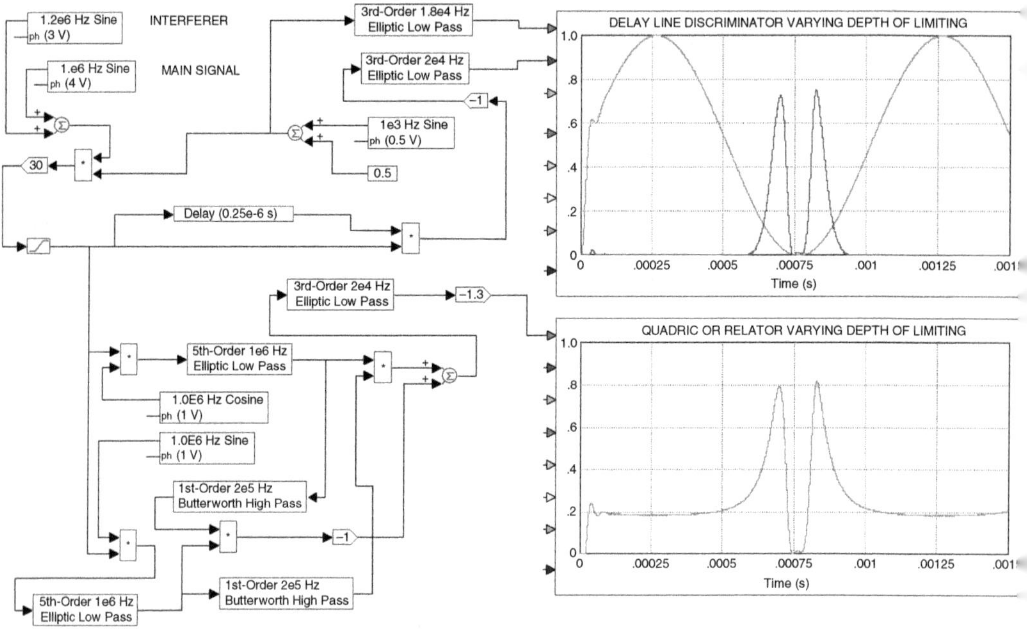

FIGURE 14.1 AM rejection with S-curve quadricorrelator.

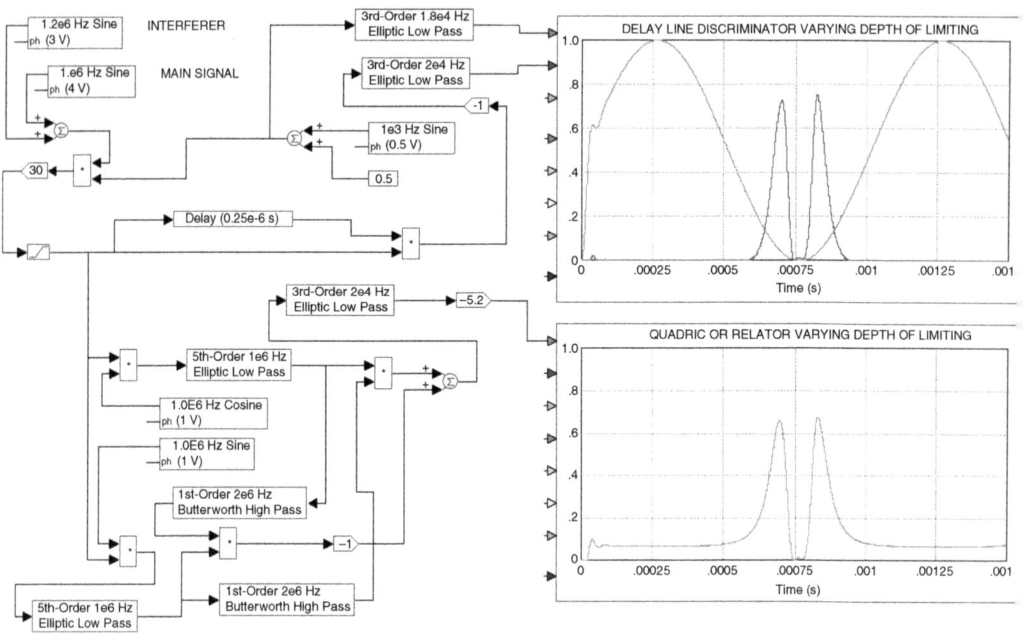

FIGURE 14.2 AM rejection by extending the S-curve width.

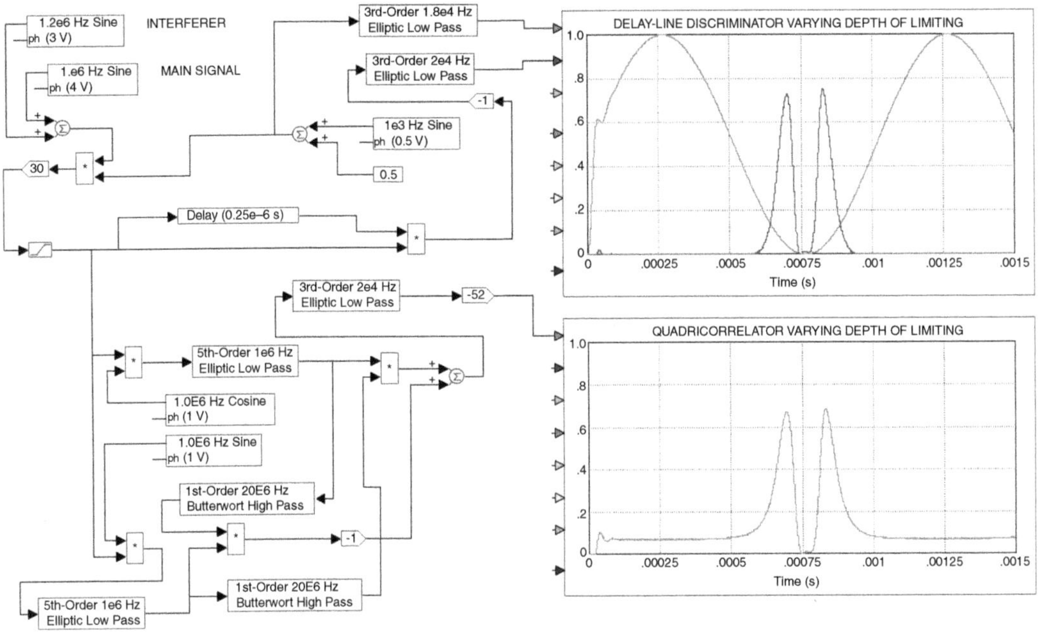

FIGURE 14.3 AM rejection does not improve appreciably when S-curve is widened between peaks.

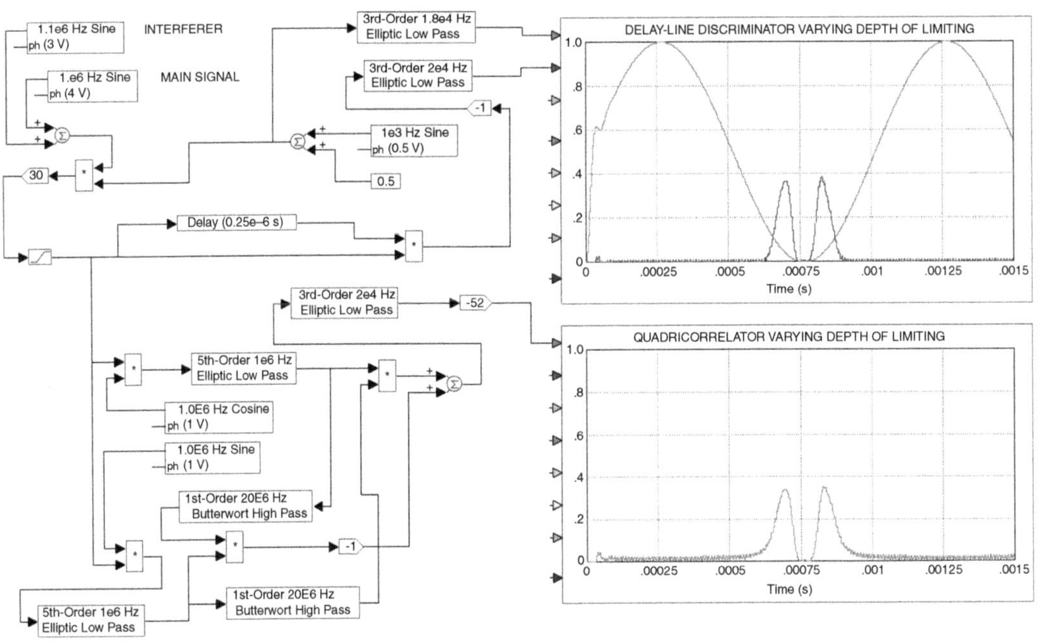

FIGURE 14.4 AM rejection improves by increasing the ratio of low-pass corners to interferer offset frequency.

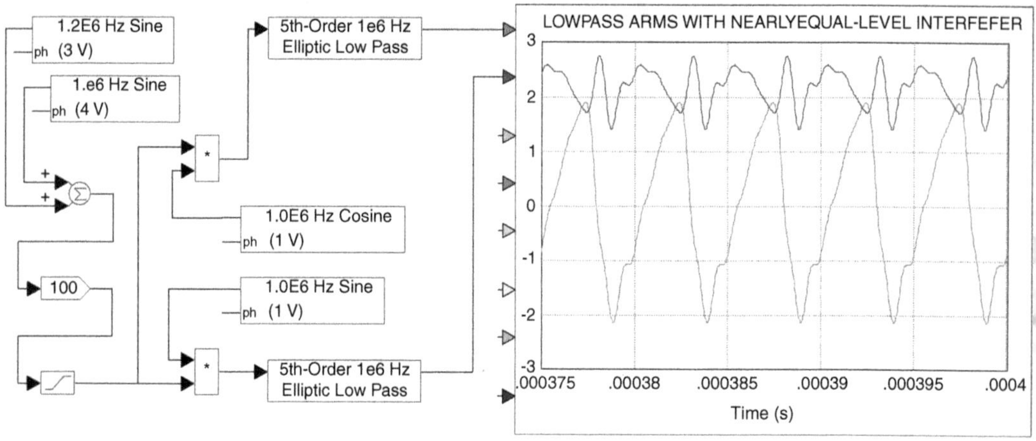

FIGURE 14.5 Quadricorrelator low-pass corner frequencies are only five times the interferer offset frequency; waveforms have kinks.

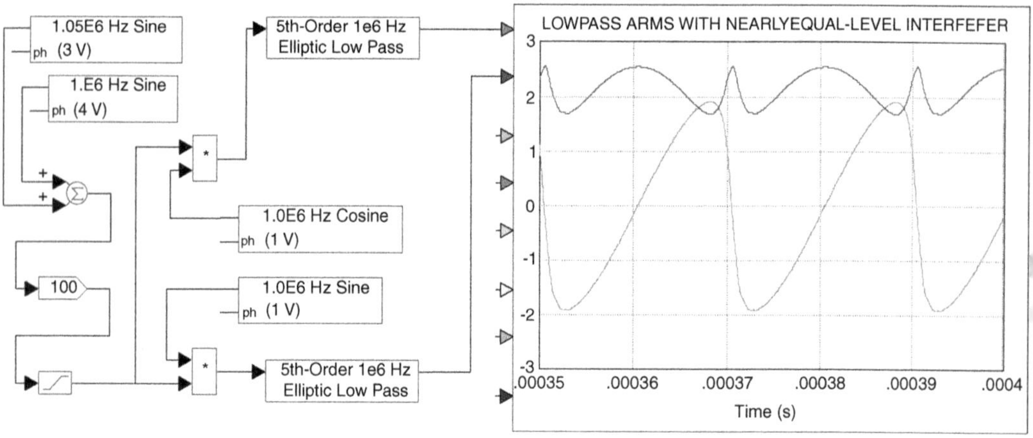

FIGURE 14.6 Quadricorrelator low-pass corner frequencies are increased to 20 times the interferer offset frequency; waveforms kinks are simpler due to greater fidelity of processing.